Decentralized Sanitation and Water Treatment

This book discusses decentralized sanitation for wastewater treatment and management in cold environments. It addresses the knowledge gap that exists between the understanding of centralized and decentralized wastewater treatment approaches. *Decentralized Sanitation and Water Treatment: Treatment in Cold Environments and Techno-Economic Aspects* covers the sustainability principles, various technologies involved, decentralized treatment in cold countries, and the economic and social feasibility of decentralized sanitation. It provides solutions for the conservation of water sources and target-oriented sanitation approaches for wastewater treatment and recycling.

Key Features

- Reviews the current status, challenges, and future perspectives of decentralized water treatments
- Discusses decentralized sanitation, water, and wastewater treatments in cold environments and Northern countries
- Focuses on interdisciplinary approaches of sustainability and the circular economy
- Covers life cycle and environmental assessment of decentralized sanitation systems
- Reviews the environmental, techno-economic, and social aspects of decentralized sanitation systems

This book is meant for professionals and researchers working on wastewater treatment, environmental engineering, and ecology.

Sustainable Industrial and Environmental Bioprocesses

This book series aims to provide a comprehensive collection of books focusing on bioprocesses in industrial and environmental biotechnology. The multidisciplinary content encompasses the chemical and biochemical engineering, industrial microbiology, and energy biosciences, all with a central theme of sustainable development and circular economy principles. The books align with the Sustainable Development Goals (SDGs) and offer state-of-the-art information and in-depth knowledge on the subject matter. The book in these series also emphasizes on the application of emerging tools, such as machine learning and artificial intelligence, for the advancement of bioprocesses. While primarily targeting academicians and researchers, the series is valuable for policy planners and industry professionals, with carefully tailored contents to cater to their specific needs and interests.

Series Editor: Dr. Ashok Pandey

Professor Ashok Pandey is currently Distinguished Scientist at the Centre for Innovation and Translational Research, CSIR-Indian Institute of Toxicology Research, Lucknow, India. His major research and technological development interests are industrial & environmental biotechnology and energy biosciences, focusing on biomass to biofuels & chemicals, waste to wealth & energy, etc. He has 16 patents, 120 books, >1000 papers and book chapters, etc. with h index of 138 and >81,000 citations. Professor Pandey is the recipient of many national and international awards and honours, which include Fellow, The World Academy of Sciences (TWAS, 2023); Fellow, Indian National Science Academy (2024); India Research Excellence Citation Award (2023), Clarivate Analytics; Highly Cited Researcher (Top 1% in the world), Clarivate Analytics (since 2018 – 2022), Rank #1 in India in Biology in India and Biochemistry and #417 in the world, Research.Com (2023), Rank #1 in India in Microbiology under Enabling and Strategic Technologies sector, Elsevier (2021), Rank #1 in India in Biotechnology and #8 in the world, Stanford University Report (2020-2023), etc.

Second and Third Generation Bioplastics: Production, Application, and Innovation
Luciana Porto de Souza Vandenberghe, Ashok Pandey, Ranjana Sirohi, Carlos Ricardo Soccol

Pharmaceuticals in Aquatic Environments: Toxicity, Monitoring, and Remediation Technologies
Vinod Kumar Garg, Ashok Pandey, Navish Kataria, Caterina Faggio

Diatoms: Ecology and Biotechnological Applications
Archana Tiwari, Ashok Pandey, Pau-Loke Show, Binod Parameswaran

Diatoms Biotechnology
Archana Tiwari, Ashok Pandey, Pau Loke Show, Binod Parameswaran

Sustainable Treatment and Management of Sewage Sludge
Mukesh Kumar Awasthi, Zengqiang Zhang, Ashok Pandey

Pharmaceuticals in Aquatic Environment: Remediation Technologies and Future Challenges
Vinod Kumar Garg, Ashok Pandey, Navish Kataria, Caterina Faggio

Biomass Hydrolyzing Enzymes: Basics, Advancements, and Applications
Reeta Rani Singhania, Anil Kumar Patel, Héctor A. Ruiz and Ashok Pandey

Decentralized Sanitation and Water Treatment: Concept and Technologies
Rajeshwar D. Tyagi, Ashok Pandey, Patrick Drogui, Bhoomika Yadav, Sridhar Pilli and Jonathan W.C. Wong

Decentralized Sanitation and Water Treatment: Treatment in Cold Environments and Techno-Economic Aspects
Rajeshwar D. Tyagi, Ashok Pandey, Patrick Drogui, Bhoomika Yadav, Sridhar Pilli and Jonathan W.C. Wong

For more information about this series, please visit: https://www.routledge.com/Sustainable-Industrial-and-Environmental-Bioprocesses/book-series/SIEB

Decentralized Sanitation and Water Treatment

Treatment in Cold Environments and Techno-Economic Aspects

Edited by

Rajeshwar D. Tyagi
BOSK Bioproducts, Québec, Canada
Dongguan University of Technology, Dongguan, China

Ashok Pandey
Centre for Innovation and Translational Research, CSIR-Indian Institute of Toxicology Research, Lucknow, India

Patrick Drogui
Institut national de la recherche scientifique Centre Eau Terre Environnment, Université du Québec, Québec, Canada

Bhoomika Yadav
Institut national de la recherche Centre Eau Terre Environnment, Université du Québec, Québec, Canada

Sridhar Pilli
National Institute of Technology, Warangal, India

Jonathan W.C. Wong
Research Center for Eco-environmental Engineering, Dongguan University of Technology, Dongguan, China

CRC Press
Taylor & Francis Group
Boca Raton London New York

CRC Press is an imprint of the
Taylor & Francis Group, an **informa** business

Design cover image: Shutterstock, 2262863233 and 2015430737

First edition published 2024
by CRC Press
2385 NW Executive Center Drive, Suite 320, Boca Raton FL 33431

and by CRC Press
4 Park Square, Milton Park, Abingdon, Oxon, OX14 4RN

CRC Press is an imprint of Taylor & Francis Group, LLC

ISBN: 978-1-032-44327-0 (hbk)
ISBN: 978-1-032-45315-6 (pbk)
ISBN: 978-1-003-37641-5 (ebk)

DOI: 10.1201/9781003376415

Typeset in Times LT Std
by KnowledgeWorks Global Ltd.

Contents

Preface

Decentralized Sanitation and Water Treatment: Treatment in Cold Environments and Techno-Economic Aspects arose from the need to develop sustainable solutions in the field of sanitation and water treatment in cold countries. This book is dedicated exclusively to the decentralized treatments in cold environments, and techno-economic and social aspects of the decentralized concepts, providing an overall evaluation of feasibility, suitability, and success of these systems. It discusses the concepts, fundamentals, and technologies of decentralized water treatment in cold countries and evaluates the environmental and economic profiles of decentralized configuration for efficient wastewater management and resource recovery and its evaluation with centralized treatment.

Water is one of the most important resources today. The inappropriate use and mismanagement of water resources has negative effects on human health, economic growth, social welfare, and ecosystems. Climate conditions are one of the crucial factors in regional water use, reuse, and management. In cold climate conditions, the issues of water treatment become more challenging, as people face significant obstacles in maintaining the natural surface and groundwater systems essential for sustainable water management. In extreme cold conditions, such as those found in northern latitudes, wastewater reuse faces unique challenges that must be addressed to ensure its feasibility and effectiveness. When applying biological treatment of sewage, more consideration should be given to the influence of low temperature on the physical and chemical properties of sewage and the activity of biological bacteria. Decentralized treatment and management systems are key considerations for treating, reusing, and recycling wastewater in extreme cold environments. Therefore, there is an urgent need for targeted research on decentralized water treatment technologies in cold regions to better understand the environmental impacts of treatment and reuse of wastewater as well as potential risks and benefits of different treatment technologies. Accordingly, this book is an attempt to compile, direct, and publicize the most relevant information about decentralized wastewater treatment in cold environments, its fundamentals, applications, and techno-economic aspects.

The contents of the book have been broadly categorized into two parts: (i) decentralized treatment in cold environments, and (ii) environmental, techno-economic, and social aspects of decentralized systems. It puts forward all the relevant topics in the form of 11 chapters and begins by introducing the water and wastewater treatment in cold environments, treatment of runoff water from deicing salt in northern countries, discusses mobile and portable treatment technologies, anaerobic wastewater treatment in extreme cold environments, and strategic issues with wastewater use in cold regions. Next, the book covers environment, health, and social acceptance of decentralized treatment. In addition, it discusses the life cycle environmental assessment of decentralized sanitation systems in developing and developed countries with technical parameters and environmental benefits. Finally, the book delves into the overall evaluation and feasibility of the decentralized system for implementation, along with challenges in the process and future perspectives. All the chapters provide state-of-the-art insights into the various technological developments in accordance with elementary research in the field of decentralized treatment, which will be ideal for future research in this field.

We expect that the book meets the requirements of the students, researchers, academicians, engineers, wastewater treatment professionals, policy makers, and rural and urban communities.

We express our deep gratitude and appreciation for the hard work and dedication of all the contributing authors in the making of this book. We thank the reviewers for their time and efforts in refining these chapters by their insightful comments. We are grateful to the Biotech Research

Society, India, for giving us the opportunity to publish this book under the agreement the Society has with Taylor & Francis. Finally, we are thankful to the team at Taylor & Francis for their cooperation in publishing this book.

Editors
Rajeshwar D. Tyagi
Ashok Pandey
Patrick Drogui
Bhoomika Yadav
Sridhar Pilli
Jonathan W.C. Wong

About the Editors

Professor Rajeshwar D. Tyagi is an internationally recognized professor of biochemical engineering and Chief Scientific Officer of BOSK Bioproducts, Québec, Canada. He has served as a professor for 36 years and as the Senior Canada Research Chair on Value Added Products from Wastes for 18 years at the University of Québec, Canada. His research on bioconversion and bioprocessing to bioproducts and activity of transfer of technology to the user's sector has received many national and international awards.

Professor Ashok Pandey is currently a Distinguished Scientist at the Centre for Innovation and Translational Research, CSIR-Indian Institute of Toxicology Research in Lucknow, India, and a Distinguished Professor, Sustainability Cluster, University of Petroleum and Energy in Dehradun, India. His major research and technological developmental interests are industrial and environmental biotechnology and energy biosciences, focusing on biomass to biofuels and chemicals, waste to wealth and energy, and more.

Professor Patrick Drogui is a professor at the Institut national de la recherche scientifique's Centre Eau Terre Environnement in Québec, Canada. He is an internationally recognized expert in electrochemical and membrane technologies and oxidative processes for different applications in water treatment from industrial wastewater to potable water.

Dr. Bhoomika Yadav completed her PhD from Institut national de la recherche scientifique's Centre Eau Terre Environnement. She has strong research skills and works in multidisciplinary areas of biotechnology, microbiology, and biochemical engineering. Her research interests include waste to value transformation, bioflocculants, wastewater treatment, bioplastics, and circular bioeconomy.

Dr. Sridhar Pilli is an Assistant Professor at the National Institute of Technology in Warangal, India. He is an expert in water and wastewater treatments, anaerobic digestion, and waste to energy.

Professor Jonathan W.C. Wong is a Distinguished Professor of the Research Center for Eco-Environmental Engineering, Dongguan University of Technology, China. Formerly, he was a professor in the Hong Kong Baptist University, Hong Kong. His major research interests are on bioconversion of organic wastes for energy and bioproducts.

Contributors

Luis Humberto Álvarez-Valencia
Departamento de Ciencias del Agronómicas y Veterinarias
Instituto Tecnológico de Sonora
Sonora, México

K. Bella
Water and Environment Division
Department of Civil Engineering
National Institute of Technology (NIT)
Warangal, India

Shraddha Chavan
Institut national de la recherche scientifique (INRS)
Centre Eau Terre Environnement (ETE)
Université du Québec
Québec, Canada

Lourdes Mariana Díaz-Tenorio
Departamento de Biotecnología y Ciencias Alimentarias
Instituto Tecnológico de Sonora
Sonora, México

Patrick Drogui
Institut national de la recherche scientifique (INRS)
Centre Eau Terre Environnement (ETE)
Université du Québec
Québec, Canada

Bhukya Gopal
Water and Environment Division
Department of Civil Engineering
National Institute of Technology (NIT)
Warangal, India

Pablo Gortáres-Moroyoqui
Departamento de Biotecnología y Ciencias Alimentarias
Instituto Tecnológico de Sonora
Sonora, México

Luis Alonso Leyva-Soto
Departamento de Biotecnología y Ciencias Alimentarias
Instituto Tecnológico de Sonora
Consejo Nacional de Ciencia y Tecnología (CONACYT)
Sonora, México

Edna Rosalba Meza-Escalante
Departmento de Ciencias del Agua y Medio Ambiente
Instituto Tecnológico de Sonora
Sonora, México

Sridhar Pilli
Water and Environment Division
Department of Civil Engineering
National Institute of Technology (NIT)
Warangal, India

Venkateswara Rao
Water and Environment Division
Department of Civil Engineering
National Institute of Technology (NIT)
Warangal, India

Ana María Rentería-Mexía
Departamento de Biotecnología y Ciencias Alimentarias
Instituto Tecnológico de Sonora
Sonora, México

Pasiano Rivas-García
Departamento de Ingeniería Química
Facultad de Ciencias Químicas
Universidad Autónoma de Nuevo León
Nuevo León, México

Denisse Serrano-Palacios
Departamento de Ciencias del Agua y Medio Ambiente
Instituto Tecnológico de Sonora
Sonora, México

Chinthala Sumanth
Water and Environment Division
Department of Civil Engineering
National Institute of Technology (NIT)
Telangana, India

Rajeshwar D. Tyagi
BOSK Bioproducts
Québec, Canada
Research Centre for Eco-Environmental Engineering, Dongguan University of Technology, Dongguan, China

Gabriela Ulloa-Mercado
Departmento de Biotecnología y Ciencias Alimentarias
Instituto Tecnológico de Sonora
Sonora, México

Zhuoyue Wang
School of Civil and Environmental Engineering
Harbin Institute of Technology
Shenzhen, China

Jonathan W.C. Wong
Research Centre for Eco-Environmental Engineering, Dongguan University of Technology, P.R. China
Hong Kong Baptist University, Hong Kong

Manojkumar Y.
Water and Environment Division
Department of Civil Engineering
National Institute of Technology (NIT)
Telangana, India

Bhoomika Yadav
Institut national de la recherche scientifique (INRS)
Centre Eau Terre Environnement (ETE)
Université du Québec
Québec, Canada

Song Yan
Institut national de la recherche scientifique (INRS)
Centre Eau Terre Environnement (ETE)
Université du Québec
Québec, Canada

Xiaolei Zhang
School of Civil and Environmental Engineering
Harbin Institute of Technology
Shenzhen, China

Part I

Decentralized Treatments in Cold Environments

1 Water and Wastewater Treatment in Cold Environments

Xiaolei Zhang[a], Zhuoyue Wang[a], Song Yan[b], Rajeshwar D. Tyagi[c,d], and Patrick Drogui[b]

[a] School of Civil and Environmental Engineering, Harbin Institute of Technology, Shenzhen, China

[b] Université du Québec, Québec, Canada

[c] Dongguan University of Technology, Dongguan, China

[d] BOSK Bioproducts, Québec, Canada

1.1 INTRODUCTION

Water is a basic human requirement, and clean drinking water is a crucial resource for human health. Water and sewage treatment capacity in cold areas is very different from treatment capacity in ordinary areas, and it is difficult to achieve the ideal effect of water treatment capacity in cold areas. Low ambient temperature is a key factor affecting water quality treatment.

Low temperatures limit the application and development of water treatment technology in cold regions. In the construction of water supply systems in cold areas and in the selection of pipeline materials, warmth and antifreeze should be considered, and certain strength and corrosion resistance should be considered when designing these systems. In the pretreatment of drinking water, more coagulants are often added to strengthen coagulation. When applying biological treatment of sewage, more consideration should be given to the influence of low temperature on the physical and chemical properties of sewage and the activity of biological bacteria. Ecological and biological technologies are frequently combined in practice to create a process that improves the efficacy of wastewater treatment and boosts the efficiency of pollutant removal. Therefore, there is an urgent need for targeted research on water treatment technologies in cold regions.

1.2 WATER SUPPLY IN NORTHERN/COLD ENVIRONMENTS

Due to the low temperatures at high latitudes, water supply is more difficult and complex in cold regions than it is in regions with normal temperatures. Permafrost can be found in some of these areas, which covers approximately 60% of Canada, as well as much of Alaska and Antarctica. Engineers consider three major issues when designing water systems for these cold regions: sources of water, pretreatment, and the effects of low temperatures on treatment facilities (Alter, 1969).

1.2.1 Sources of Water

Water supply sources are frequently numerous in regions with normal temperatures. However, in freezing northern regions, water sources are often scarce, and there are fewer lakes and rivers to maintain appreciable flows all year. Smaller lakes and rivers are easily frozen in the winter.

DOI: 10.1201/9781003376415-2

Permafrost limits the flow and use of groundwater, making it harder to discover suitable water sources in colder regions. When winter frosts reach the top of the permafrost layer, the water above it becomes useless.

Although water under the permafrost layer can provide a continual supply of water, these sources are extremely hidden and difficult to identify. In addition, this type of water is expensive to produce and frequently includes significant quantities of iron and manganese (Hopkins, 1955).

1.2.2 Pretreatment

Water quality in cold regions varies with the changing of seasons, thus it requires some pretreatment. Water turbidity levels are often surpassed in water sources influenced by surface runoff, such as in the Yukon and Northwest Territories in Canada. The Slave and Hay rivers bring enormous amounts of silt into Great Slave Lake from early May until late August, coloring the lake dark or green (Dickens, 1959). Therefore, the water body must be thoroughly filtered before usage. Because of the peculiar local soil, rich in humus, in the northeast of China, a large amount of humic acid-like substances enter the reservoirs with surface runoff during precipitation, resulting in a high humic content in the reservoir water. As a result, a large amount of coagulant is needed for treatment. Water pretreatment in low temperatures requires longer operation times for mixing and aeration, which raises the cost of water supply.

1.2.3 Impact of Low Temperature on Facilities

The design of water supply systems must consider freeze protection due to the colder temperatures in the northern regions. In places where outdoor temperatures are below –20°C, insulation measures need to be taken for outdoor water meter wells, valve wells and pump combiners.

Water supply treatment equipment and processes in cold areas must be properly cased, insulated and frost-protected. Significant condensation difficulties might arise in the construction of treatment facilities. Therefore, treatment facility design and construction should take care to insulate and protect against moisture, as well as provide ventilation during operation (Alter, 1969).

1.3 DRINKING WATER SUPPLY TO INDIGENOUS COMMUNITIES

Undoubtedly, the rapid growth in global population will encourage more people to immigrate to regions in the Arctic and subarctic. Modern water distribution systems are currently needed because of these changes, and this requirement will only get worse over time.

1.3.1 Common Cold Area Water Distribution Systems

For instance, consider the North Pole. The main challenge for water supply engineers in the Arctic is maintaining liquid water across a distribution system. Basically, just two of the several methods created for this purpose are in use. The system is either built such that the water is evacuated from the system before it freezes, or sufficient heat is applied to the water to avoid freezing (Murphy and Hartman, 1969).

Intermittent pumping: Throughout the summer, the system is run similarly to any other regular distribution system. However, water cannot be left in the lines during the winter. As a result, the water is sent via the system on demand or according to a set timetable, and each customer fills a storage tank inside his home. This only happens in cases when full management of the water is practicable, such as in extremely tiny settlements and other low-use settings.

Conventional system with deep burial and/or insulation: This type of system is workable in many far northern locales outside of permafrost zones. Heat must be provided, the pipe must be buried

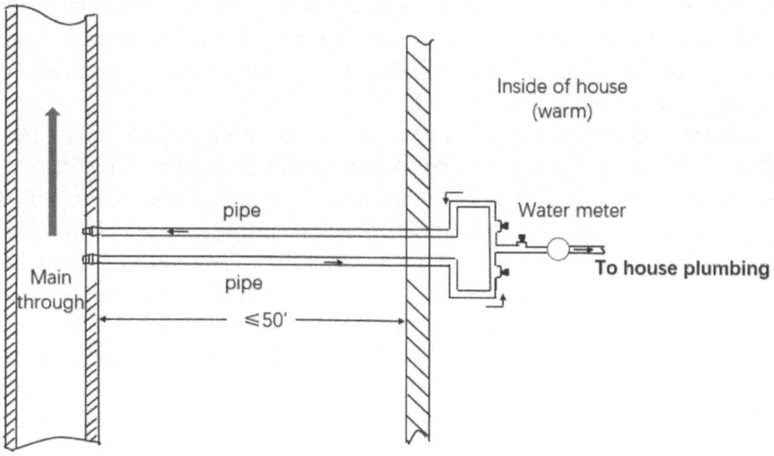

FIGURE 1.1 Schematic diagram of the single-main recirculating system.

deeper and more insulation must be added when the distribution system is lengthened or water demand decreases.

Single-main recirculating system: This method of water distribution consists of one water main through (Murphy and Hartman, 1969). The water is continuously pumped during periods of extreme cold or in areas of permafrost. The flow loop is from the treatment plant or primary pumping station around the loop and either back into the same pumping station or into another station. Heat is added to the water, when necessary, at the pumphouse. The problem of keeping the water in the service connections in motion without reverting to bleeding or individual pump was solved by the development of a device that has become known in the North as the "pitorifice", a word derived from both pitot tube and orifice.

Figure 1.1 is a schematic diagram of the single-main recirculating system.

As shown in Figure 1.2, the single-main recirculating system consists of one or more uninterrupted loops originating at a recirculation facility and returning to that point without any branch loops. This layout eliminates dead ends and related freezing problems and requires the minimum amount of piping as compared to other circulation methods. Normally, water is pumped out at

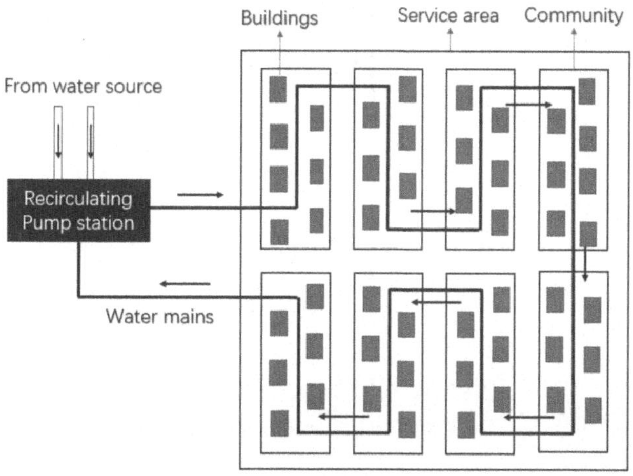

FIGURE 1.2 Layout and location of mains for single-pipe recirculation.

between 39°F and 45°F and returns at 33°F to 39°F. If possible, the mains will be located at the rear of buildings, rather than in the streets. This provides greater thermal protection (since snow is not typically removed), results in less risk of damage to manholes and other appurtenances and allows shorter, less-costly service connections to the buildings.

Alternative systems: Conventional water pipe mains with no recirculation are possible in very special situations. In these cases, there may be enough flow in the system so that return loops are not needed if a high-volume consumer is placed at the ends of the main line. A dual pipe system relies on a large diameter supply line and a small diameter return line placed side by side in a utilidor or a pre-insulated conduit, as shown in Figure 1.3. The return line is sized to maintain the desired flow in the system. The service lines are taken off the main and returned to the smaller diameter return line. The pressure differential between the delivery and the return line must be sufficient to induce circulation in the service loop. This system is complex, and control mechanisms tend to be elaborate. For example, varying consumption in different locations can result in stationary water in areas at certain times. Thermostatically controlled solenoid valves are used at regular intervals to overcome this problem.

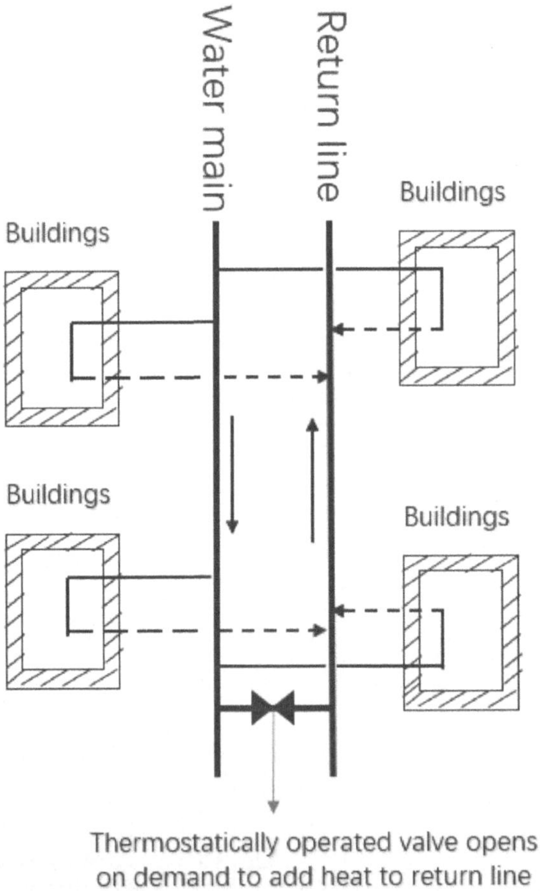

FIGURE 1.3 Schematic diagram of a dual pipe system.

1.3.2 Plumbing Considerations

The design of service lines must not only prevent freezing of the contained water but also consider the effects of permafrost thawing, frost heaving and differential settlements between the pipes and the building. Circulation in these service connections is the recommended method to prevent freezing. Heat tapes are also commonly installed as a backup. The two most common methods for circulation are either a small pump inside the building or the use of pit orifices. There will be a small velocity-induced pressure differential between the upstream and downstream orifices, which in turn induces flow in the service loop. If the service lines are longer than 80 ft, the pressure differential in the main is not sufficient and the pit orifice will not function. A small circulation pump must be added in the building for longer distances (Obara et al. 2011.

1.3.2.1 Water Pipe Insulation Material

The discussion that follows is intended to give some indication of the performance of these materials in the Arctic and subarctic.

1. Copper. Type K copper is commonly used for service lines because it can be thawed by using electrical resistance, whereas plastic pipes cannot be thawed in this manner.
2. Ductile iron. Ductile iron pipe can take some shock loadings and is slightly flexible. It is a heavy, durable pipe often used in rocky areas or where adequate pipe bedding materials are not available. It can be thawed with electrical resistance.
3. Steel pipe. Steel pipe is lighter, more flexible and more corrosion resistant than ductile iron. It can also be thawed using electrical resistance.
4. High density polyethylene (PE). Pipe of this material is very flexible and impact resistant, with a high coefficient of expansion and contraction, high corrosion resistance and a smooth interior, but it cannot be threaded.
5. Others. Polyvinyl chloride (PVC), acrylonitrile-butadiene-styrene (ABS), asbestos-cement (AC), reinforced concrete and even wood stave piping (Obara et al. 2011).

1.3.2.2 Insulation Material for the Outer Layer of the Water Pipe

1. Polyurethane foam. Polyurethane foam is a thermal insulation material used in directly buried pipelines. It is characterized by small heat loss, good pressure resistance, good thermal insulation performance, good anticorrosion insulation performance, long service life, convenient construction and installation, small footprint and low engineering cost.
2. Advanced rubber and plastic. It is used for the same pipeline with thin thermal insulation thickness and less consumption. It is a green, environmentally friendly clean-type insulation material, with less waste in construction and harmless to health.
3. Others. Glass wool, composite silicate, rock wool pipe, etc.

1.3.2.3 Backup Freeze Protection

The following are reserve systems to either prevent freezing if circulation stops in the lines or to thaw the system if freezing occurs. The procedures apply to both main lines and service piping. These include electrical heat or steam tracing to prevent freezing or to thaw, and electrical resistance heating and steam or hot water to thaw systems.

1. Heat trace systems. A heat trace system is the standard backup used in most piped water distribution systems.
2. Steam or hot water thawing. This system uses a source of steam or hot water introduced under pressure into the frozen pipe via a suitable hose or tube to thaw out the pipe.

3. Service line thawing. Small-diameter service lines of any material may be quickly thawed by pushing a flexible 0.5 inch, or smaller, plastic tube into the frozen pipe while pumping warm water into the tube.
4. Electrical resistance thawing. The thawing of metal pipes using electricity is fairly common.

1.4 DRINKING WATER TREATMENT IN INDIGENOUS COMMUNITIES

Depending on the kind and quality of the raw water, several treatment methods are used to produce drinking water. Aeration, flocculation, sedimentation, filtration and disinfection are common treatment techniques that guarantee water is of acceptable quality for ingestion (Figure 1.4). Indigenous communities in cold regions primarily treat their water using the following technical procedures (Craun, 1988).

1.4.1 Pre-oxidation

Chemical pre-oxidation is a pretreatment step that adds oxidants to the beginning of the water treatment process to improve the treatment impact. Pre-oxidants like ozone and potassium permanganate are frequently utilized. Its major goals are to eliminate odor, regulate chlorination disinfection byproducts, control oxidation coagulation and remove iron and manganese.

Researchers employed combined ozone/hydrogen peroxide technology to increase the breakdown efficiency of extremely stable organic contaminants in water to address the issue of poor effluent quality in cold locations (Jasim et al., 2012). The rate of ozone breakdown and the formation of hydroxyl radicals are accelerated by a little amount of hydrogen peroxide added to the ozone oxidation process. In comparison to ozone molecules, hydroxyl radicals have a greater oxidation, react with organic materials more quickly, and are less temperature dependent (von Gunten, 2003). Therefore, low temperature conditions are ideal for the efficient breakdown of the majority of organic contaminants in water. Based on this, the loaded honeycomb ceramic Fe–Mn–Cu–K and the heterogeneous catalytic honeycomb ceramic worked together with hydrogen peroxide to catalyze the decomposition of ozone to produce hydroxyl radicals with better efficiency. The porosity

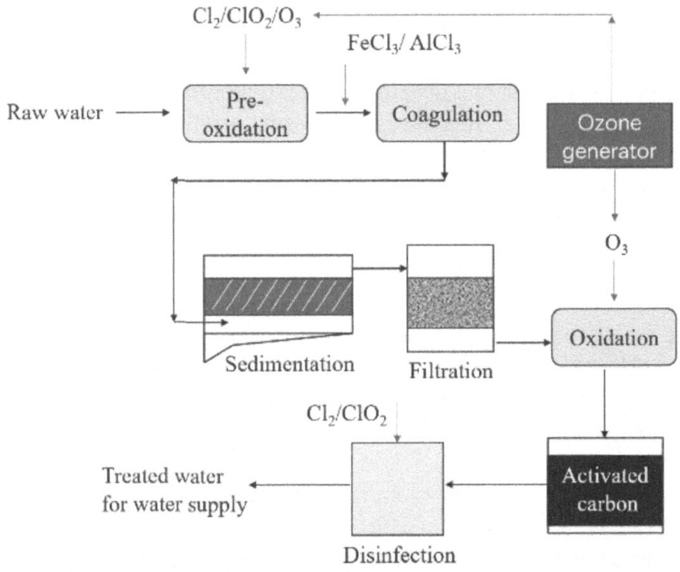

FIGURE 1.4 The normal procedure of water treatment in cold regions.

of honeycomb ceramics facilitates the dispersion of ozone, accelerates the decomposition rate of oxidants in the ozone/hydrogen peroxide system, and also promotes the generation of hydroxyl radicals in the system. Pre-coagulation ozonation has previously been shown to improve following drinking water treatment and enhance the quality of final water (Jasim et al., 2008). Additionally, it reduces the precursors of trihalomethanes (THMs) and assists in the elimination of turbidity and particulates.

1.4.2 Coagulation

After the pre-oxidation stage, coagulation, which also includes flocculation, is a crucial phase in the conventional drinking water treatment process. Adding coagulant to raw water destroys the stable state of colloidal particles, allowing the particles to easily adsorb and contact each other, causing the water to condense. Under certain hydraulic conditions, the generated flocculant materials due to the collision and aggregation between the colloidal particles and other particles, are easily to separate from the water by flocculation. However, as the power needed for the traditional coagulation technology and the dose of coagulant increases, the effluent quality gets worse because of the increase in water's viscosity coefficient in very cold environments. These problems can be solved by using polymer and high-efficiency coagulant aids in combination with the administration of monitoring technologies.

The coagulation effect is greatly enhanced when polyaluminum sulfide chloride (PASC) is used as a coagulant (Zhonglian, Baoyu, and Qinyan, 2010). The most effective method for removing turbidity and pollution is to combine polyaluminum chloride (PAC) and PAFC with activated silicic acid in cold water (Yang et al., 2012).

In order to strengthen the coagulation process, the second-phase water supply project of the Harbin Mopanshan water treatment plant employs rotating grid technology, inserting a tiny pore-size grid on the stirring paddle of the conventional mechanical flocculation tank. The rotating grid solves the problem of little disturbance of the flow in the area between the paddles of the traditional stirring paddle, allowing the energy to be distributed more evenly in the water, thus promoting more adequate flocculation. The spinning grid's mesh aperture and wire diameter can be adjusted, and the speed of the grid can be modified by changing the stirring paddle's speed. This technology is suitable for the reconstruction and construction of large-scale water purification plants.

1.4.3 Filtration

Filtration generally refers to the process of intercepting suspended solids and other impurities in water through the surface or filter layer of a filtration medium. The biological efficiency of the filter media is compromised due to the low temperature in the alpine environment. This issue can be resolved using the air flotation, membrane separation and micro-flocculation contact filtration technologies (Craun, 1988).

More and more membrane technologies are being used in recent years to treat low-temperature, low-turbidity water. Membranes can filter away water odor, chromaticity, and precursors to disinfection byproducts. As shown in Table 1.1, there are four groups of membrane processes: microfiltration (MF), ultrafiltration (UF), nanofiltation (NF), and reverse osmosis (RO) (Smith, 1996). In addition, water quality can be improved by using immersion membrane technology to treat low-temperature, low-turbidity water (Davis and Cornwell, 1998).

Micro-flocculation contact filtration is when coagulation and filtration are finished in a process. The addition of coagulant is done directly into the filter material layer for filtration, skipping the steps of precipitation. A micro-flocculation-enhanced filtration process was used to treat water with low temperature and turbidity. The results showed that the process has a strong ability to remove turbidity, organic matter and particulate matter.

TABLE 1.1
Characteristics of Four Membrane Processes

	MF	UF	NF	RO
Pore size	0.1–0.4 µm	0.01–0.2 µm	100–400 Da	20–100 Da
Configuration	Hollow fiber membranes	Hollow fiber membranes	Sheets of membranes wound around a perforated core tube	Sheets of membranes wound around a perforated core tube
Removal Objective	Colloidal contaminants including microbial pathogens such as Giardia lamblia and Cryptosporidium parvum from source waters	Colloidal contaminants including microbial pathogens such as Giardia lamblia and Cryptosporidium parvum from source waters	Dissolved contaminants such as NOM, which imparts a yellow-brown color to water	Dissolved salts, such as Ca^{2+} and Mg^{2+}

1.5 DECENTRALIZED WASTEWATER TREATMENT IN COLD ENVIRONMENTS

For urban areas with a dense population and well-developed drainage systems, centralized treatment is appropriate. Decentralized wastewater treatment is appropriate for regions with dispersed settlements, underdeveloped economies, lagging pipe network coverage, etc. Decentralized wastewater treatment technology could achieve water cycle and balance *in situ* treatment and *in situ* reuse (Yu et al., 2015). It can effectively treat wastewater for landscape and agricultural irrigation, industrial use, etc. (Lautze et al., 2014; González et al., 2015; Bajpai, Katoch, and Chaturvedi, 2019).

The two main categories of decentralized wastewater treatment are biological treatment technology and ecological treatment technology. Biological treatment technologies include membrane bioreactors (MBR), activated sludge techniques, biogas digesters and biofilm wastewater treatment processes, such as bio-filters, bio-rotaries, biological contact oxidation and bio-fluidized beds (Kobayashi et al., 2020). Ecological treatment technologies include constructed wetlands (CWs), land infiltration systems, stabilization ponds and ecological filters. On the basis of a moderate financial investment, ecological and biological technologies are frequently combined in practice to create a process that improves the efficacy of wastewater treatment and boosts the efficiency of pollutant removal.

Currently, combining wastewater treatment processes has been carried out in order to complement the advantages of several single processes, and the combination of biological methods with electrochemical systems has proved to be the most promising for application, including overflow electrochemical membrane bioreactor (EMBR) (Zhou et al., 2015), microbial fuel cells (MFCs) (Wang et al., 2012) and advanced septic reactor (ASR) (Jorsaraei, Gougol, and Van Lier, 2014).

While various technologies have shown promising results in terms of removal of contaminants, their applicability in cold environments is a topic that is frequently disregarded. Only a few studies have focused on developing treatment technologies that can work efficiently in low temperatures (Figure 1.5) (Varma et al., 2021). Decentralized wastewater treatment technologies rely heavily on microorganisms and plants, but lower temperatures reduce the activity of microorganisms and plants, thus limiting the performance of the treatment process.

1.5.1 Constructed Wetland Treatment

Constructed wetland treatment is an artificial ecosystem that simulates a natural wetland, formed by a composite ecosystem of filler substrate–aquatic plants–microorganisms (Dordio and Carvalho, 2013; Wu et al., 2015) through substrate filtration, adsorption, sedimentation, ion exchange, plant

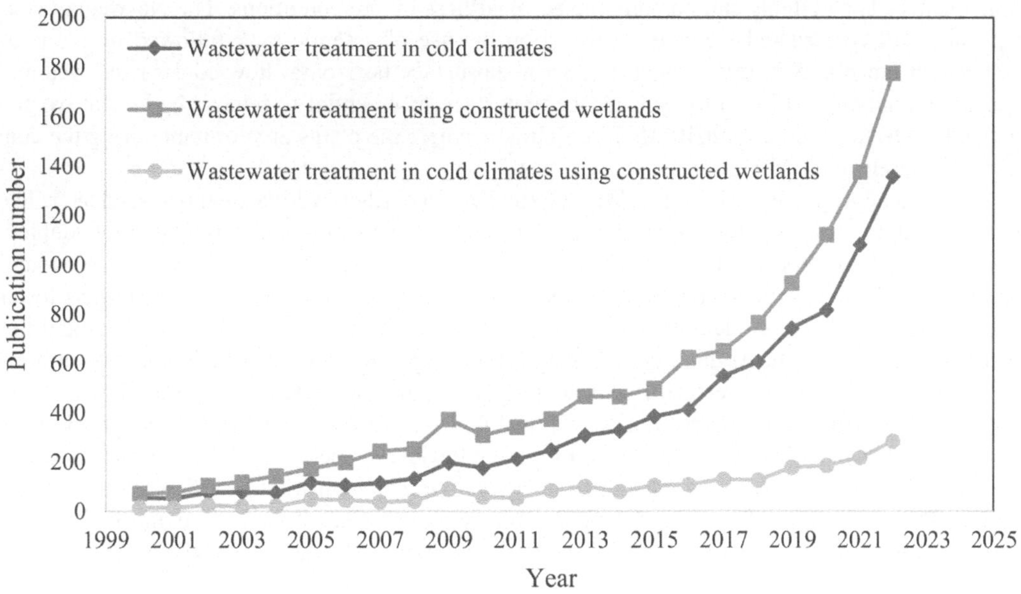

FIGURE 1.5 Frequency of publications on wastewater treatment in cold environment.

absorption and microbial decomposition (Ye et al., 2012; Chen, Zuo, and Guo, 2012). Constructed wetlands are divided into vertical subsurface wetlands (VSW), horizontal subsurface flow wetlands (HSFW) and surface flow wetlands (SFW) according to the flow of wastewater (Zhang et al., 2012). The advantages of which are ecological sustainability, low energy consumption (Hench et al., 2003) and landscape effects; the disadvantages are low pollutant removal loads and the need for large areas of land.

It is worth noting that temperature is a limiting factor in constructed wetlands. Low temperature had the most antagonistic effect on the performance of the CWs because freezing ambient temperature caused ice formation, hydraulic imperfections, malfunctioning of biotic and abiotic components, etc. Varma et al. (2021) studied around 20 different scenarios in order to find suitable remedies to overcome the challenges faced in cold climates, such as incorporating greenhouse structure, providing insulating materials, bio-augmentation, etc. According to several studies (Liu et al. 2019; Xu et al. 2018), greenhouse construction led to a 20% increase in removal of total nitrogen (TN) and chemical oxygen demand (COD), while plant collocation accounted for up to an 18% increase in the removal of COD. All of the artificial aeration, insulation and bio-augmentation enhanced the performance of the CWs in low temperatures.

Reuter, Djohan, and Goldman (1992) were among the first to investigate the use of wetlands in cold regions to purify micro-polluted rivers. The research shows that wetland plants still have some restoration capacity in cold areas. However, when the temperature is low, wetland plants become less effective in cleaning water bodies, and the leaves may even wilt or die, causing secondary pollution to the water body. To address this shortcoming, cold-resistant plants are used instead of conventional plants to enhance the growth and development of plants at low temperatures. In addition, the restoration effect is fundamental to effectively improve the ability of cold-resistant plants to remove nitrogen and phosphorus under low-temperature stress, which enables plants to maintain a certain ability to purify water in cold environments (Wang et al., 2017).

In 1991, the first subsurface flow constructed wetland for treatment of domestic wastewater was built in Norway. Today, this method is rapidly becoming a popular method for wastewater treatment in rural Norway. This is due to excellent performance, even during winter, and low maintenance. The systems can be constructed regardless of site conditions. The Norwegian concept for small constructed wetlands is based on the use of a septic tank followed by an aerobic vertical down-flow biofilter succeeded by a subsurface horizontal-flow-constructed wetland. The aerobic biofilter, prior to the subsurface flow stage, is essential to remove biological oxygen demand (BOD) and achieve nitrification in a climate where the plants are dormant during the cold season (Jenssen et al., 2005).

The artificial wetlands in Shenyang World Expo Park in northern China are introduced later. The high-efficiency composite flow artificial wetland sewage treatment technology is mainly adopted. The compound flow constructed wetland technology is to complete the combination of horizontal subsurface flow and vertical flow in the same wetland unit, which occupies less land, has lower operating costs and is more flexible in operation. This process has a unique innovation in the water distribution method of the wetland system, that is, the composite water distribution method of horizontal submerged flow and vertical rise is adopted. Compared with other composite flow wetland systems, the combination of horizontal submerged flow and vertical flow can be completed in the same wetland unit, so that the wetland system is more flexible in operation. It can alternate between dry and wet in different growing seasons of plants, maximize the use of the efficiency of the wetland, make full use of the function of the plant root system and, at the same time, most of the suspended matter and shed biofilm are deposited into the wetland filler (Kataki et al., 2021).

1.5.2 Stabilization Ponds

Stabilization ponds are artificially constructed ponds with embankments and permeable layers that efficiently remove organic matter and pathogens from wastewater mainly via microbial degradation, organic matter adsorption, organic particle settling and filtration (by anaerobic microorganisms, aerobic microorganisms, planktonic organisms and aquatic plants) (Li et al., 2019). The removal rate of BOD_5 in stabilized ponds is frequently above 90%, but its removal of nitrogen (N) and phosphorus (P), which cause eutrophication in water bodies, is less effective. Thus, the screening of plants in the system needs to be enhanced and dynamically controlled in subsequent work (Sun, Tian, and Li, 2014; Yu et al., 2015). Sun et al. (2014) found that the attapulgite combined with stabilization pond could improve sewage treatment effect in cold regions.

In Norway, sewage from a community consisting of 160 people is treated using a pond/reed bed system. Treatment performance during the first five years has been high and unaffected by harsh winter conditions. Research carried out at Solborg Camphill Village in Norway showed that wastewater stabilization ponds aerated by flow forms provided an effective means of aerobic treatment in temperatures of −10°C and with 50 cm of ice cover on the ponds (Jørgensen et al., 2005).

1.5.3 Soil Infiltration Technology

Soil infiltration technology is a land treatment system in which wastewater is artificially controlled and dispensed into a soil layer with a certain structure (Figure 1.6), at a specific depth from the ground and with good diffusion properties, using soil–plant–animal–microbial physicochemical properties to achieve the required treatment and use. By laying pipe above an anaerobic tank and laying straw on the surface of cultivated soil, the innovated construction of soil leachate bed for sewage treatment in cold regions was found.

1.5.4 Moving Bed Biofilm Reactor

The moving bed biofilm reactor (MBBR) uses the metabolic activity of various microorganisms to improve the pollutants' removal by biofilm, thus achieving the purpose of wastewater purification. Temperature is one of the key factors affecting the efficiency of the MBBR process. Nitrifying bacteria and denitrifying bacteria are both medium temperature bacteria, the most suitable temperature

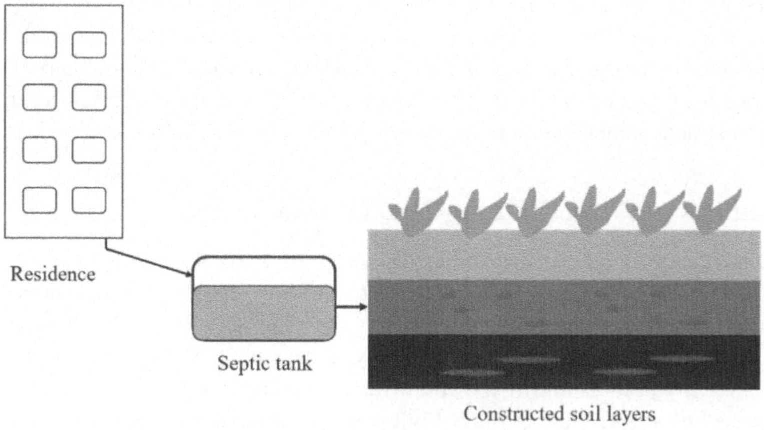

FIGURE 1.6 Soil infiltration treatment system.

range is 20–35°C, too high or too low a temperature would inhibit their the growth and reproduction, thus decreasing the denitrification and deteriorating the quality of effluent (Hoang et al., 2014).

1.5.5 Lifting Applications in Cold Regions

To promote the application of decentralized wastewater treatment process in cold regions, many researchers have upgraded the existing process. Cao, and Di (2022) and Li et al. (2023) designed a decentralized courtyard ecological sewage and wastewater treatment system with solar thermal insulation greenhouse. To promote application in plateau severely cold agricultural and pastoral areas, Chenget al. (2018) changed the heat storage structures that are equipped with heat coils at the bottom of the septic tank reactor. Liao et al. (2017) designed a heat pipe MBR domestic wastewater treatment system for use in cold regions.

In the Arctic, permafrost and surface bedrock complicate installation, and due to the cold climate, it is necessary to insulate the tanks. In areas with permafrost or deep frost penetrations, the septic tanks may even have to be installed in a heated area under the house or in a heated shed. Composting and anaerobic digestion are treatment methods that could be well suited for black water. Both processes may be more challenging to run in cold climates than in more temperate zones, thus sufficient insulation and heating of reactors are needed (Yao et al., 2020).

Cold climates will cause a series of problems and challenges in the processes of wastewater treatment. This could, for instance, include freezing in treatment of wastewater and reduction of biological activity. Therefore, treatment of wastewater is often inadequate or completely lacking in cold regions. Studies of possibilities to take advantage of the cold climate when designing treatment methods for cold regions should be undertaken. The examples in this chapter provide the basis for renewed interest in the potential of decentralized wastewater treatment systems in cold climates.

1.6 CHALLENGES IN WATER AND WASTEWATER TREATMENT IN COLD ENVIRONMENTS

1.6.1 Challenges of Water Source and Sewage Quality

In the cold regions of high latitude, the phenomenon of ice and ice melting often occurs in rivers and lakes. The seasonal change of water source is usually accompanied by change in the physical state and quality of water, which has an impact on the access of water supply sources. Safe water conveyance is a major concern for many countries when building water conservation projects, particularly during the ice age. At the same time, the water quality of sewage must be correctly grasped as a precondition of the design of the sewage treatment process. Generally, pollutant levels in sewage are typically correlated with local climatic factors and lifestyle choices. On a global scale, residents of high latitude regions typically eat meat, which leads to a higher content of oil in sewage, which has made sewage treatment somewhat more challenging (Alves et al., 2009).

1.6.2 Challenges of Water Collection and Delivery Systems

The collection, transportation, utilization or discharge of water supply and drainage all depend on unobstructed pipeline systems. However, due to the low temperature in cold areas, the conveying water or sewage may be frozen, as well as the pipe network and structures, which may bring great harm to normal production and safety. Steam heat, electric heat tracing, and other methods are currently used as antifreeze and insulation measures in conventional water systems, however these techniques frequently have high energy requirements and significant limitations. The impact of frozen soil on pipeline laying is another issue that cannot be disregarded in some extremely cold regions.

1.6.3 Challenges of Treatment Structures and Facilities

Based on operation practices, the construction of primary sedimentation tank, aeration tank and secondary sedimentation tank exposure in air is safe, and insulation measures such as filling soil and surrounding the exposed part can be taken. However, it is advisable to build freezable facilities indoors, such as screens and the sand discharge part of the sand sink. If the water produces large-scale freezing, there will be a risk of damage to screens and other mechanical facilities. Furthermore, for a sewage treatment plant located in a permafrost region, the blower room is supposed to configure the air preheater. That raises the cost of site construction, infrastructure materials and manpower, resulting in more challenges to be overcome.

1.6.4 Challenges of Water and Wastewater Treatment Technology

Generally speaking, water and wastewater treatment technologies commonly used include physical, chemical and biological methods, and their treatment processes are affected by natural environmental factors, such as temperature, to varying degrees.

Physical methods often involve the role of sedimentation removal. According to the Stokes formula, the settling rate of particles is negatively correlated with the viscosity coefficient of water. When the temperature drops, the settling of particulate matter is hampered, and the treatment effect is worsened due to an increase in the viscosity coefficient of water. According to pertinent data, primary sedimentation tanks are usually designed with a retention time of 1.5–1.6 h based on the average flow rate. The residence time needed to achieve the same efficiency for wastewater at 10°C is 1.38 times longer than it is at 20°C. The effects are particularly significant in colder climates.

Chemical methods contain coagulation, neutralization, redox, etc. Similar to the physical method, low temperatures can inhibit some of these processes, especially when it comes to certain reversible reactions. For example, low temperatures can prevent the hydrolysis process when aluminum-containing flocculants are used in a treatment system, which reduces the efficacy of the treatment.

Biological processes are also involved in water and wastewater treatment, in which microorganisms play a significant part in purification or as a means of reducing pollutants. Temperature is a crucial ecological factor that affects the development and survival of microorganisms and has a significant impact on human growth, reproduction and metabolism. It has been demonstrated that low temperatures can obviously inhibit the nitrification reaction. The nitrifying bacteria's biological activity nearly ceases at temperatures below 4°C (Xu et al., 2014). Thus, cold climatic conditions will cause a series of problems and challenges in the biological processes of wastewater treatment.

1.7 SUSTAINABLE WATER TREATMENT TECHNOLOGIES FOR NORTHERN ENVIRONMENTS

The sustainability of water resources is becoming a major concern as a result of the increasing global population and consequent demands on water. Thus, it is important to address water issues in a more integrated and innovative way. In cold climates, traditional techniques are only partially effective, which calls for the development and application of sustainable and low-energy technologies in these areas.

1.7.1 Screening and Construction of Cryogenic Engineered Bacteria

To reduce engineering investment and operation costs, it is wise to look for ways to increase the biological efficiency of low-temperature wastewater treatment. It is concluded that pro-cooling microorganisms can sustain and become dominant at ambient temperatures below 10°C, based on the unique characteristics of their membrane proteins, lipids and genetic response to heat transfer (Dev et al., 2019). And the psychrophilic anaerobic digestion (AD) operation has proven to be an economically sustainable alternative process in cold regions, in which the biogas produced can be used for combustion heat generation. At the same time, the process does not require heating and cooling systems, nor does it need to cover insulation walls for structures such as digestion vessels, which makes the construction and operation of the plant less costly in terms of capital. Previously, SK-4 was isolated from sediments from Lake Nagaike in Antarctica to assess its potential for use in activated sludge systems (Tsuji et al., 2013). Therefore, the cultivation and domestication of cold-adapted microorganisms for water treatment in cold regions have great development prospects.

1.7.2 Microbial Fuel Cells

In conventional processes, effluent is usually used (reclaimed water) to discharge directly, while sustainable processes seek to recover all useful resources. In this regard, wastewater is comparable to a resource or energy carrier. Microbial fuel cells (MFCs), as a newly developed technology, can effectively and sustainably produce bioenergy, such as biological hydrogen and electricity, by using organic matter in wastewater as a carbon source to finish the removal of pollutants during the energy conversion process (Chen et al., 2022). The operation of MFC systems has been shown to be less temperature-sensitive than conventional biological treatment, and it can be used in areas without adequate electrical infrastructure, suggesting that it has development potential.

1.7.3 Discharge Plasma Technology

In recent years, an emerging discharge plasma technology, an advanced oxidation technology (AOPs), has often been applied to the deep treatment of water. It has been proved that nonthermal plasma technology (NTP) can produce ·OH, O_3, H_2O_2 and other strong oxidizing substances directly while maintaining low temperature (Ma et al., 2020), without producing any toxic residues. And the NTP process requires shorter running time and performs with higher process efficiency (Chandana et al., 2018; Murugesan et al., 2020). However, it is crucial to remember that some plasma technologies have poor economic conditions due to excessive energy consumption and running costs. As a result, various novel combination processes have been developed, such as the synergy of discharge plasma technology with persulfate activated oxidation. The combination of physical, chemical, and electrical effects that constitute the process's bactericidal effect ensures a deep removal of organic matter with low energy cost, which is crucial for the cost management of realistic wastewater treatment projects.

1.8 FUTURE RESEARCH

At present, there are still certain issues for the biological treatment of water and wastewater to be resolved in cold regions. Low temperature will have an impact on the community structure of sludge microorganisms as well as inhibit their activity, which will eventually result in poor effluent quality. To overcome the existing problems, further experimental research can be carried out, mainly from the following aspects.

1. Carry out in-depth experimental research on the existing mature treatment process and modify the sewage treatment process under low-temperature conditions and regulate the operating conditions. The newly built sewage treatment plant uses and transforms the biofilm process with better adaptability (Li et al., 2020), and proposes new reactors that are more suitable for treatment in low temperature (Ai et al., 2023). At the same time, the mechanism of various enhanced treatment processes under low temperature conditions needs to be deeply studied to provide theoretical support for the improvement and popularization of wastewater-enhanced treatment processes in cold environments.
2. Isolate preponderant psychrotrophs and psychrophiles for cultivation and domestication (Xu et al., 2018). A large number of studies have shown that the isolated psychrotolerant bacteria can effectively degrade COD and various pollutants. Because the metabolic mechanism of psychrotolerant microorganisms enables them to grow well at low temperatures, it is very promising to apply psychrotolerant microorganisms to degrade organic pollutants in psychrotolerant environments. Further research can be carried out from the cold adaptation mechanism of psychrotolerant bacteria at the molecular level (Furhan, 2020), and nitrogen and phosphorus removal bacteria with high activity under low temperature conditions can be constructed by genetic engineering technology.

1.9 CONCLUSIONS AND PERSPECTIVES

Due to the characteristics of high latitude and low temperature in cold regions, water supply sources, pretreatment measures and the impact of low temperature should be considered when designing a water supply system. In cold regions, proper water sources are often lacking, and raw water needs more careful pretreatment. For the facilities of the water supply system, the design needs to consider such factors as thermal insulation and freezing lines.

The primary difficulty faced by the water supply system in cold regions is how to inexpensively solve the problem of water freezing in the pipeline. In order to prevent the freezing problems, recirculation is normally applied at a large or small scale. In addition, the heat is supplied when it is in extreme cold conditions. Moreover, the water supply to the buildings is also kept in flow loop, which can indirectly heat the water as it is warmed in the building. In fact, the most effective way for avoiding the freezing problems is to develop new types of pipe material that is heatproof. Another option is to use pipe with an inner pipe for water supply and casing pipe for warm water circulating to keep the water from freezing. Currently, some water supply pipeline in cold regions employs insulation material to wrap the pipes for preventing them from freezing.

The water quality in alpine regions is often poor due to characteristic low oxygen. Therefore, compared with the conventional treatment process for drinking water (coagulation, sedimentation, filtration, disinfection), it is very necessary to add some pretreatment, to strengthen the conventional process or advanced treatment process to improve the quality and safety of drinking water.

Wastewater treatment, whether it is by a physical method, chemical method or biological method, will be affected to varying degrees by low temperature. In order to solve the problems associated with low temperature, this chapter introduces some low energy consumption and sustainable technologies that may be suitable for application in cold regions. Among them, the screening and construction of low-temperature engineering bacteria have good application prospects. If the short-term application effect is pursued, the existing mature treatment process can be deeply tested and studied, the sewage treatment process under low-temperature conditions can be modified and the operating conditions can be adjusted.

REFERENCES

Adibi, M. M. (2000). "Cold Load Pickup," in Power System Restoration: Methodologies & Implementation Strategies, *IEEE*, 2000, 105–107, doi: 10.1109/9780470545607.ch14.

Ai, Shengshu, Linzhu Du, Zebing Nie, Wenai Liu, Hua Kang, Fan Wang, and Dejun Bian. 2023. 'Characterization of a novel micro-pressure double-cycle reactor for low temperature municipal wastewater treatment', *Environmental Technology*, 44: 394–406.

Alter, Amos J. 1969. *Water supply in cold regions* (US Army Materiel Command, Terrestrial Sciences Center, Cold Regions Research ...).

Alves, M. M., M. A. Pereira, D. Z. Sousa, A. J. Cavaleiro, M. Picavet, H. Smidt, and A. J. M. Stams. 2009. 'Waste lipids to energy: How to optimize methane production from long-chain fatty acids (LCFA)', *Microbial Biotechnology*, 2: 538–50.

Bajpai, Mukul, Surjit Singh Katoch, and Naveen Kumar Chaturvedi. 2019. 'Comparative study on decentralized treatment technologies for sewage and graywater reuse–a review', *Water Science and Technology*, 80: 2091–106.

Chandana, L., C. J. Sangeetha, T. Shashidhar, and C. Subrahmanyam. 2018. 'Non-thermal atmospheric pressure plasma jet for the bacterial inactivation in an aqueous medium', *Science of the Total Environment*, 640: 493–500.

Cheng, Shikun, Li Zifu, Sayed Mohammad Nazim Uddin, Mang Heinz-Peter, Zhou Xiaoqin, Zhang Jian, Zheng Lei, and Zhang Lingling. 2018. 'Toilet revolution in China', *Journal of Environmental Management*, 216, 347–356.

Chen, T. M., C. Zou, J. J. Pan, M. S. Wang, L. Qiao, F. H. Wang, Q. Zhao, H. Y. Cheng, C. Ding, and Y. Yuan. 2022. 'Mapping research on microbial fuel cells in wastewater treatment: A co-citation analysis', *Processes*, 10: 16.

Chen, Yu-quan, Zhuo Zuo, and Xiao Guo. 2012. 'Application of a stereo constructed wetland mode to the treatment of slightly polluted source Water', *Meteorological and Environmental Research*, 3: 63.

Craun, Gunther F. 1988. 'Surface water supplies and health', *Journal-American Water Works Association*, 80: 40–52.

Davis, Mackenzie L., and David A. Cornwell. 1998. *'Environmental Engineering', Editorial Mc Graww Hill.*

Dev, S., S. Saha, M. B. Kurade, E. S. Salama, M. M. El-Dalatony, G. S. Ha, S. W. Chang, and B. H. Jeon. 2019. 'Perspective on anaerobic digestion for biomethanation in cold environments', *Renewable & Sustainable Energy Reviews*, 103: 85–95.

Dickens, H. B. 1959. 'Water supply and sewage disposal in permafrost areas of northern Canada', *Polar Record*, 9: 421–32.

Dordio, Ana V., and Alfredo Jorge Palace Carvalho. 2013. 'Organic xenobiotics removal in constructed wetlands, with emphasis on the importance of the support matrix', *Journal of Hazardous Materials*, 252: 272–92.

Furhan, Junaid. 2020. 'Adaptation, production, and biotechnological potential of cold-adapted proteases from psychrophiles and psychrotrophs: Recent overview', *Journal, Genetic Engineering & Biotechnology*, 18: 36.

González, O., B. Bayarri, J. Aceña, S. Pérez, and D. Barceló. 2015. 'Treatment technologies for wastewater reuse: Fate of contaminants of emerging concern', In: Fatta-Kassinos, D., Dionysiou, D., Kümmerer, K. (eds) *Advanced Treatment Technologies for Urban Wastewater Reuse. The Handbook of Environmental Chemistry*, vol 45. Springer, Cham, 5–37.

Hench, Keith R., Gary K. Bissonnette, Alan J. Sexstone, Jerry G. Coleman, Keith Garbutt, and Jeffrey G. Skousen. 2003. 'Fate of physical, chemical, and microbial contaminants in domestic wastewater following treatment by small constructed wetlands', *Water Research*, 37: 921–27.

Hoang, Valerie, Robert Delatolla, Edith Laflamme, and Alain Gadbois. 2014. 'An investigation of moving bed biofilm reactor nitrification during long-term exposure to cold Temperatures', *Water Environment Research*, 86: 36–42.

Hopkins, David Moody. 1955. *Permafrost and ground water in Alaska* (US Government Printing Office).

Jasim, S. Y., S. Ndiongue, O. Alshikh, and A. T. Jamal. 2012. 'Impact of ozone and hydrogen peroxide vs. UV and hydrogen peroxide on chlorine residual', Ozone*: Science & Engineering*, 34: 16–25.

Jasim, S. Y., S. Ndiongue, B. Johnson, L. Schweitzer, and D. Borikar. 2008. 'The effect of ozone on cold water coagulation', *Ozone: Science & Engineering*, 30: 27–33.

Jenssen, Petter D, Trond Mæhlum, Tore Krogstad, and Lasse Vråle. 2005. 'High performance constructed wetlands for cold climates', *Journal of Environmental Science and Health*, 40: 1343–53.

Jørgensen S.E., Löffler H., Rast W., and Straškraba M. 2005. 'Chapter 2 – Lake and reservoir water uses and abuses', In: Jørgensen, S.E., Löffler, H., Rast, W., Straškraba, M. (eds.) *Developments in Water Science*, 54. Elsevier, 43–106.

Jorsaraei, Abbas, Mahdi Gougol, and Jules B. Van Lier. 2014. 'A cost effective method for decentralized sewage treatment', *Process Safety and Environmental Protection*, 92: 815–21.

Kataki, S., Chatterjee S., Vairale M.G., Sharma S., Dwivedi S.K., and Gupta D.K. 2021. 'Constructed wetland, an eco-technology for wastewater treatment: A review on various aspects of microbial fuel cell integration, low temperature strategies and life cycle impact of the technology', *Renewable and Sustainable Energy Reviews*, 148: 111261.

Kobayashi, Yumi, Nicholas J. Ashbolt, Evan G. R. Davies, and Yang Liu. 2020. 'Life cycle assessment of decentralized greywater treatment systems with reuse at different scales in cold regions', *Environment International*, 134: 105215.

Lautze, Jonathan, Emilie Stander, Pay Drechsel, Allegra K. da Silva, and Bernard Keraita. 2014. 'Global experiences in water reuse', CGIAR Research Program on Water, Land and Ecosystems (WLE). International Water Management Institute (IWMI), Colombo, Sri Lanka, 31.

Li, Dan, Xinze Wang, Lina Chi, and Jie Wang. 2019. 'The design and operation of subsurface wastewater infiltration systems for domestic wastewater', *Water Environment Research*, 91: 843–54.

Li, Keqian, Ren Wuang, Wang Qianle, Lu Xu, Shi Xuan, Xue Bai, Xin Jin, Wang Xiaochang, and Jin Pengkang. 2023. 'Onsite treatment of decentralized rural greywater by ecological seepage well (ESW)', *Journal of Cleaner Production*, 393: 136180.

Li, Ren, Li-Fang Yu, Xing-Xiu Zhang, Zi-Cheng Dai, Si-Si Hua, and Dang-Cong Peng. 2020. 'Adaptability of nitrifying biofilm systems to low temperature: MBBR and IFAS', *Huan Jing Ke Xue= Huanjing Kexue*, 41: 3691–98.

Liao, Xiaoping, Caixiang Zhang, Yanxin Wang, and Mi Tang. 2017. 'The abiotic degradation of methyl parathion in anoxic sulfur-containing system mediated by natural organic matter', *Chemosphere*, 176: 288–95.

Liu, Miaomiao, Masaru K. Nobu, Jia Ren, Xiaowei Jin, Gang Hong, and Hong Yao. 2019. 'Bacterial compositions in inhalable particulate matters from indoor and outdoor wastewater treatment processes', *Journal of Hazardous Materials*, 385: 121515.

Ma, S., K. Kim, S. Chun, S. Y. Moon, and Y. Hong. 2020. 'Plasma-assisted advanced oxidation process by a multi-hole dielectric barrier discharge in water and its application to wastewater treatment', *Chemosphere*, 243: 9.

Murugesan, P., V. E. Monica, J. A. Moses, and C. Anandharamakrishnan. 2020. 'Water decontamination using non-thermal plasma: Concepts, applications, and prospects', *Journal of Environmental Chemical Engineering*, 8: 23.

Obara, Shin'ya, Watanabe, Seizi, and Rengarajan, Balaji. 2011. 'Operation planning of an independent microgrid for cold regions by the distribution of fuel cells and water electrolyzers using a genetic algorithm', *International Journal of Hydrogen Energy*, 36: 14295–14308.

Reuter, John E, Tjut Djohan, and Charles R Goldman. 1992. 'The use of wetlands for nutrient removal from surface runoff in a cold climate region of California—Results from a newly constructed wetland at Lake Tahoe', *Journal of Environmental Management*, 36: 35–53.

Sun, Nan, Weiwei Tian, and Chenyang Li. 2014. 'Sewage treatment effect in cold rural region by attapulgite combined with stabilization pond', *Transactions of the Chinese Society of Agricultural Engineering*, 30: 209–15.

Symeon, Christodoulou, Michalis Fragiadakis, Agathoklis Agathokleous, and Savvas Xanthos 2017. *Urban Water Distribution Networks*. 1st Edition. Butterworth-Heinemann.

Tsuji, M., Y. Yokota, K. Shimohara, S. Kudoh, and T. Hoshino. 2013. 'An application of wastewater treatment in a cold environment and stable lipase production of Antarctic basidiomycetous yeast *Mrakia blollopis*', *PloS One*, 8: 7.

Varma, Mahesh, Ashok Kumar Gupta, Partha Sarathi Ghosal, and Abhradeep Majumder. 2021. 'A review on performance of constructed wetlands in tropical and cold climate: Insights of mechanism, role of influencing factors, and system modification in low temperature', *Science of the Total Environment*, 755: 142540.

von Gunten, U. 2003. 'Ozonation of drinking water: Part I. Oxidation kinetics and product formation', *Water Research*, 37: 1443–67.

Wang, Mo, Dong Qing Zhang, Jian Wen Dong, and Soon Keat Tan. 2017. 'Constructed wetlands for wastewater treatment in cold climate—A review', *Journal of Environmental Sciences*, 57: 293–311.

Wang, Yong-Peng, Xian-Wei Liu, Wen-Wei Li, Feng Li, Yun-Kun Wang, Guo-Ping Sheng, Raymond J Zeng, and Han-Qing Yu. 2012. 'A microbial fuel cell–membrane bioreactor integrated system for cost-effective wastewater treatment', *Applied Energy*, 98: 230–35.

Wu, Haiming, Jian Zhang, Huu Hao Ngo, Wenshan Guo, Zhen Hu, Shuang Liang, Jinlin Fan, and Hai Liu. 2015. 'A review on the sustainability of constructed wetlands for wastewater treatment: Design and operation', *Bioresource Technology*, 175: 594–601.

Xu, Guangsu, Han, Yunping, Lin Li, and Liu, Junxin. 2018. 'Characterization and source analysis of indoor/outdoor culturable airborne bacteria in a municipal wastewater treatment plant', *Journal of Environmental Sciences*, 74: 71–78.

Xu, G. R., C. Jia, Z. Zhang, and Y. L. Jiang. 2014. 'Enhanced physicochemical-biological sewage treatment process in cold regions', *Water Science and Technology*, 70: 1456–64.

Xu, Z. Z., Y. Ben, Z. L. Chen, A. X. Jiang, J. M. Shen, and X. Y. Han. 2018. 'Application and microbial ecology of psychrotrophs in domestic wastewater treatment at low temperature', *Chemosphere*, 191: 946–53.

Yang, Z. L., B. Y. Gao, Y. Wang, Y. Q. Zhao, and Q. Y. Yue. 2012. 'Fractionation of residual Al in natural water treatment from reservoir with poly-aluminum-silicate-chloride (PASiC): Effect of OH/Al, Si/Al molar ratios and initial pH', *Journal of Environmental Sciences*, 24: 1908–16.

Yao Yao, Gordon Huang, Chunjiang An, Xiujuan Chen, Peng Zhang, Xiaying Xin, Jian Shen, and Joy Agnew. 2020. 'Anaerobic digestion of livestock manure in cold regions: Technological advancements and global impacts', *Renewable and Sustainable Energy Reviews*, 119: 109494.

Ye, Jianfeng, Liang Wang, Dan Li, Wei Han, and Chao Ye. 2012. 'Vertical oxygen distribution trend and oxygen source analysis for vertical-flow constructed wetlands treating domestic wastewater', *Ecological Engineering*, 41: 8–12.

Yu, Yanyan, Haibo Li, Hong Wang, Xin Wang, Xianchao Ji, and Yulong Duan. 2015. "Status of Studies on the Ecological Disposal of the Decentralized Wastewater Treatment Technologies." In *2015 2nd International Workshop on Materials Engineering and Computer Sciences*, 251–54. Atlantis Press.

Zhang, Ting, Dong Xu, Feng He, Yongyuan Zhang, and Zhenbin Wu. 2012. 'Application of constructed wetland for water pollution control in China during 1990–2010', *Ecological Engineering*, 47: 189–97.

Zhonglian, Yang, Gao Baoyu, and Yue Qinyan. 2010. 'Coagulation performance and residual aluminum speciation of $Al_2(SO_4)_3$ and polyaluminum chloride (PAC) in Yellow River water treatment', *Chemical Engineering Journal*, 165: 122–32.

Zhou, Guowang, Yuhong Zhou, Guoqiang Zhou, Lian Lu, Xiankai Wan, and Huixiang Shi. 2015. 'Assessment of a novel overflow-type electrochemical membrane bioreactor (EMBR) for wastewater treatment, energy recovery and membrane fouling mitigation', *Bioresource Technology*, 196: 648–55.

2 Treatment of Runoff Water from Deicing Salt for Road Safety in Cold Climate Conditions

Song Yan[a], *Patrick Drogui*[a], *Rajeshwar D. Tyagi*[b,c], *and Jonathan W.C. Wong*[b]
[a] Université du Québec, Québec, Canada
[b] Dongguan University of Technology, Dongguan, China
[c] BOSK Bioproducts, Québec, Canada

2.1 INTRODUCTION

The salinization of freshwater bodies due to the use of road deicing salts has become a growing concern in recent years. Road salts, such as sodium chloride, magnesium chloride, and calcium chloride, are commonly used in many regions of the world to improve winter driving conditions. However, the salts are often washed off the roads during rainfall or snowmelt and end up in nearby freshwater systems. This can result in high concentrations of salts in rivers, lakes, and groundwater, which can have lasting ecological effects and can even contaminate drinking water supplies.

The negative impacts of road salts on species at all trophic levels, physical and chemical properties of freshwater systems, and drinking water supplies highlight the urgent need for effective mitigation measures.

In colder climate regions of North America, deicing salts applied to roads have been identified as a major contributor to local and regional contamination. Total salt use in North America has increased dramatically, mostly due to increased road salt application (Findlay & Kelly, 2011).

Deicing chemicals work by lowering the freezing point of water. The most common deicers are inorganic chloride-based salts, i.e. sodium chloride (NaCl), magnesium chloride ($MgCl_2$), and calcium chloride ($CaCl_2$). Other trace elements may include phosphorous, sulfur, nitrogen, copper, and zinc (Soleimanifar, 2019). After application, road salts dissolve and enter freshwater ecosystems through saline overland flow generated by snowmelt or rain and through groundwater sources. Road salt is a persistent freshwater contaminant and freshwater communities can be exposed for long periods of time, depending on the dynamics of surface and groundwaters (Findlay & Kelly, 2011; Ledford, Lautz, & Stella, 2016; Mason, Norton, Fernandez, & Katz, 1999).

Current levels of salt contamination highlighting spatial patterns, evidence for retention, and potential mechanisms are causing slow transport through ecosystems. Direct and indirect effects of these quantities on a variety of organisms, habitats, and ecological processes have been considered (Findlay & Kelly, 2011; Hintz & Relyea, 2019). Widespread use of salts as deicing agents on roads has been perceived as a significant source of environmental and economic damage. Not only near-road and short-term effects were was investigated, but also the indirect and long-term road salt effects on ecosystems (Findlay & Kelly, 2011).

In this chapter, the impact of deicing salt on urban lakes and the environment is reviewed. Conventional techniques used for treatment and management of runoff water from deicing salt are

DOI: 10.1201/9781003376415-3

highlighted. In addition, recovery and reuse of deicing salt as better alternatives are illustrated, and the future perspectives are described, all in an attempt toward reaching a balance between keeping roads safe for humans during the winter months and protecting freshwater ecosystems from road salt salinization.

2.2 IMPACT OF DEICING SALT ON URBAN LAKES AND ENVIRONMENT

Road deicing salt affects many species across all trophic levels at environmentally relevant concentrations. All types of freshwater ecosystems—lakes, rivers, streams, and wetlands—are affected by human-induced salinization. The rising salinity of fresh waters from the use of road deicing salts has received increasing attention (Schuler et al., 2019).

2.2.1 Urban Lakes

Compared to rivers, streams, and wetlands, fewer studies have examined the impacts of road salts on lake communities. In a natural lake, road salt concentrations of 217–445 mg Cl^-/L have been associated with changes in the species composition of phytoplankton, zooplankton, and benthic communities toward more salt-tolerant species (Bridgeman et al., 2000; Judd et al., 2005). From the same lake, the authors suggested that sediment cores revealed that following road salt contamination, one genus of phytoplankton common to mesotrophic and brackish waters (*Asterionella*) became far more prevalent, while large-bodied zooplankton grazers like *Daphnia* declined. As a result, there was reduced grazing pressure on phytoplankton communities (Judd et al., 2005), which may lead to elevated phytoplankton abundance.

Using paleolimnological analyses, Valleau, Paterson, and Smol (2020) studied the long-term effects of Cl^- from road salt runoff on shallow, soft-water lakes within the southern Canadian Shield. Specifically, they examined Cladocera subfossils in dated sediment cores from six lakes within the Muskoka River watershed of south-central Ontario, Canada. Long-term changes were compared in cladoceran assemblages in the five lakes with the highest regional Cl^- concentrations (32.8–90.9 mg/L) with a reference lake (0.91 mg/L). Overall, biological changes in the impacted lakes were best explained by changing Cl^- concentrations associated with road salt application. Among lakes, the severity of impact may have been indirectly mediated by differences in food quantity through its effect on Cl^- toxicity (Valleau, Paterson, & Smol, 2020).

In lake ecosystems with low calcium (Ca) concentrations, road salt might lead to negative effects on zooplankton species' richness and abundance at much lower Cl^- concentrations than systems with higher Ca concentrations (Jensen, Meland, Schartau, & Walseng, 2014). In two low-Ca European lakes recovering from acidification, Jensen et al. (2014) found that the recovery of the zooplankton community in the lake with a road salt concentration of 6.9 mg Cl^-/L (average Ca concentration of 0.9 mg/L) was slower than recovery in the lake with a concentration of 3.2 mg Cl^-/L (average Ca concentration of 1.1 mg/L). However, Jensen et al. (2014) presented no significant indication that Cl^- concentration was solely responsible, and added that the low Ca concentration in the lake with greater road salt contamination may have also been responsible for the slower recovery of the zooplankton community (Ashforth & Yan, 2008; Hessen, Alstad, & Skardal, 2000).

Combining the assessment of the community- and species-level effects of road salts in lake systems, Hintz and Relyea (2019) expected dramatic shifts in diversity and food web dynamics with elevated road salt concentrations. Since road salt can trigger trophic cascades that cause algal blooms, shading from these phytoplankton blooms could exacerbate any direct toxic effects of road salts on benthic communities through a reduction of benthic primary production (i.e. food resources). Thus, a dense saline layer at the bottom of lakes combined with altered lake hydrology could compound cascading effects on benthic communities.

Changes in the abundance and diversity of these resources could have bottom-up effects that reduce fish recruitment. A reduction in the recruitment of some fish species could then negatively affect the piscivores in a lake food web (Hintz & Relyea, 2019).

2.2.2 Streams and Rivers

Using NaCl, Cañedo-Argüelles et al. (2016) found salinity stress occurring in an invertebrate predator at 7,000 μS/cm was insufficient to trigger a trophic cascade in artificial streams. In a separate experiment, salt pulses increased the abundance of salt-tolerant invertebrate species in streams without changing species richness (Cañedo-Argüelles et al., 2014). Among basal trophic levels, an *in situ* experiment showed that the standing crop and diversity of autotrophs can be reduced significantly in streams at concentrations >600 mg Cl^-/L (Dickmann & Gochnauer, 1978). Field surveys appear to yield dramatically different results than experimental studies of stream communities. For example, an analysis of 41 streams across a salinity gradient found that the diversity of diatoms was significantly reduced at 35 mg Cl^-/L (Porter-Goff et al., 2013). Another field study of ten watersheds found that macroinvertebrate community composition changes occur at salt concentrations as low as 50–90 mg Cl^-/L (Wallace & Biastoch, 2016). Road salt can also alter the diversity of higher-level consumers in streams.

In the Australian Alps, the ecological effects of even small salinity increases from deicing may be different than in North America and Europe because of (1) differences in the evolutionary history, and (2) areas with deicing in Australia are not located in urbanized landscapes where deicing has been largely studied elsewhere. The salinity increases attributable to deicing in Australia and the effects of this increase in salinity to stream macroinvertebrates was determined (Shenton et al., 2022). Increased salt concentrations (as measured by continuous measurements of electrical conductivity (EC) and periodic measurements of chloride concentrations) were observed in streams near two Australian ski resorts, during the snow seasons (June to September) of 2016 to 2018. The maximum EC observed in streams in salted sites near Perisher, New South Wales, was 390 μS cm^{-1} compared with a maximum of 26.5 μS cm^{-1} at unsalted sites. Lower EC values (i.e. maximum 61.1 μS cm^{-1}) and short durations of salinity increases in streams near Falls Creek, Victoria, were not expected to cause an adverse biological response. Salt storage in the landscape was evident at salted sites near Perisher where EC was above background levels during periods of the year when no salt was applied to roads. Stream macroinvertebrate community composition differed at sites receiving runoff from road salting activities near Perisher. Abundances of Oligochaeta (worms) (up to 11-fold), Dugesiidae (flat worms) (up to fourfold), and Aphroteniinae (chironomids) (up to 14-fold) increased, whereas Leptophlebiidae (mayflies) decreased by up to 100% compared with non-salted sites. The taxa that were less abundant where deicing salts were present tended to be the same taxa that toxicity testing revealed to be relatively salt-sensitive species. The results demonstrated a causal link between deicing salts, elevated stream salinity, and altered macroinvertebrate community composition in streams that received runoff from road deicing activity in the Australian Alps (Shenton et al., 2022).

Shenton et al. (2022) conducted a study to generate knowledge on the environmental effects of deicing salt, particularly sodium chloride (NaCl), on water quality in Minnesota, especially the Twin Cities Metropolitan Area (TCMA). The Mississippi River receives substantial sodium chloride inputs from the Minnesota River and wastewater treatment plants as it passes through the TCMA. In addition, road salt applications in the TCMA use about 350,000 short tons of NaCl every year. A chloride budget at the scale of the TCMA and on individual sub-watersheds in the TCMA indicates that about 70% of the road salt applied in the TCMA is not carried away by the Mississippi River. Rates of seasonal road salt use are correlated with snowfall, road miles, and population. Salinity in TCMA lakes increases in winter and decreases in summer. Ionic composition of dissolved substances in lakes of the TCMA suggests unnaturally high sodium and chloride concentrations compared to lakes and other water bodies in the midwestern United States. Data indicate a rising trend in urban lake water salinity over the last 30 years. Shallow groundwater in the TCMA, especially near major roadways, has started to show increasing chloride concentrations. Salinity trends in lakes and shallow aquifers of the TCMA are of concern (Stefan, Novotny, Sander, & Mohseni, 2008).

2.2.3 Wetlands

Road salt can also have dramatic effects in permanent and seasonal wetland communities (including stormwater and treatment ponds). At the basal level of the community, road salt contamination at relatively low concentrations can reduce the richness, evenness, and coverage of native plants in natural wetlands (i.e. 54 mg Cl^-/L; Richburg, Patterson, & Lowenstein, 2001). Tadpole grazing combined with 645 mg Cl^-/L can reduce microbial respiration by as much as 24%, which may reduce energy flow provided by detrital resources in experimental pond food webs (Van Meter, Swan, & Trossen, 2012).

More research is needed to determine how cascading responses are modulated by abiotic conditions in wetlands (e.g. pH, dissolved oxygen, temperature).

A constructed natural ecosystem involving phytoremediation could serve as a buffer for such pollution sources. Constructed wetlands have been previously used to purify point sources of road runoff before it enters the environment. Retention of contaminants does occur, due to sedimentation and adsorption. Studies of the effects of first flush phenomena and deicing on the retention capacity and the outflow pollutant concentrations have shown quite variable results, with potential mobilization of copper, zinc, and lead as a result of contact with salt water.

There was a dramatic effect of deicing salts on the concentrations of Cu, Zn, Cd, and Ni in the effluent leaving the system. For Cu, the concentrations even became higher than they had ever been in the road water. It is advised to let the road water bypass the facility during deicing periods (Tromp, Lima, Barendregt, & Verhoeven, 2012).

2.2.4 Other Environments

Several roads in Kosciuszko National Park (KNP) in New South Wales, Australia, are kept open each winter with the use of snow ploughs and the deicing salts sodium chloride and calcium chloride. Alpine Bogs and Fens, which are endangered Sphagnum cristatum bog communities of high conservation significance, receive saline runoff from deiced roads in KNP. Chloride deicing salts raise the osmotic potential of soil water and degrade soil physical properties, and both sodium and chloride ions are toxic to alpine flora. However, the impact of deicing salts on alpine environments has not been previously studied in Australia. To establish salt loads and residence times in Alpine Bogs, bog groundwater electrical conductivity (EC) was measured weekly using a network of piezometers in 16 roadside bogs during winter and spring 2010. A treatment of 7.5 kg of NaCl and 1 GBq of tritium (3H) was diluted into 100 L of water and applied to two bogs to establish if deicing salts moved slower through the bogs than groundwater. Monitoring revealed a significant flux of salt through roadside bogs. ECs were raised significantly in salted bogs with levels highest closest to the road, 3,520 µS/cm, and dropping to 100 µS/cm at a distance of 120 m from the road. The 3H injection suggested NaCl moved slower than groundwater, and salts had a shorter residence time in wetter bogs. The results indicated that Alpine Bog hydrology was significantly altered by current snow clearing practices. Peak ECs were significantly higher than the 30 µS/cm of undisturbed bogs and might potentially retard bog flora vigor (Hocking et al., 2011).

The use of NaCl for road deicing has been found to release both particulate and dissolved pollutants from the road surface, particularly in regions with frequent and prolonged seasonal frosts such as northern countries. For example, in Sweden, significant amounts of trace elements are mobilized in roadside soils every year due to the use of NaCl as a deicing agent. This is a danger to the drinking water sources located in shallow areas called eskers. Additionally, in sewer systems where large amounts of particulate pollutants accumulate on road surfaces or in gully pots, deicing can also result in the release of pollutants into the surrounding environment. To address these concerns, national and regional governments in the Netherlands have established standards for receiving surface waters and groundwater bodies with respect to point sources of stormwater runoff (Szklarek, Górecka, & Wojtal-Frankiewicz, 2022).

Long-term winter application of road salt increases the annual mean concentrations of Cl in rivers and lakes, due to Cl entering groundwater. The second step was a cause-effect analysis of the impact of NaCl on the abiotic processes in soil and water, and on the biotic response to chloride exposure. Chlorides appear to decrease the biodiversity of aquatic animals and plants but favor the growth of phytoplankton, especially cyanobacteria. Moreover, Cl reduces the self-purification processes of water by decreasing nutrient accumulation in macrophytes, decreasing the denitrification rate, and reducing organic matter decomposition. It indicated that a system-based approach integrating engineering knowledge with an understanding of biological and hydrological processes is necessary to determine solutions for reducing environmental risks from road salt use (Szklarek, Górecka, & Wojtal-Frankiewicz, 2022).

By far the largest data sets were obtained for sodium chloride. Magnesium and potassium chloride appear to be more toxic than sodium chloride for all organisms investigated. Plankton and invertebrates appear to be more sensitive to calcium than sodium chloride, whereas the reverse appears to be true for fish. Chloride salts from road salt are toxic in the environment although in limited geographic areas where large amounts of salt are applied to highways or near leaking road salt depots. As chloride concentration gradually increases, there are unknown changes in composition and standing stock as a result of biological interactions. Because road salt has the potential to alter natural communities (like elevated phosphorous), elevated concentrations should be viewed as being potentially environmentally harmful (Evans & Frick, 2001).

The respiratory activity of microorganisms in sewage sludge is affected by several physicochemical factors, and a considerable amount of information concerning the influence of individual factors has been recorded. However, very little is known about their effects in combination. In the recent past, the concentration of chlorides entering wastewater has increased greatly and the major source is sodium chloride, which is used in large quantities to deice roads in winter. The effects of various concentrations of sodium chloride at different temperatures on the respiratory activity of the microorganisms have been examined. It is found that low temperature (6°C) and high concentrations of NaCl (1 M) had drastic effects on the metabolism of the microorganisms in activated sludge; the acclimation phase of activated sludge is 117 hours. Not only is their respiratory activity strongly depressed but also the length of their acclimation period is greatly magnified (Vítězová, Nitayapat, & Vítěz, 2018).

To inform mitigation efforts, the transport of chloride in surface waters of a metro-area watershed (Lake McCarrons) was studied in this project to characterize chloride transport by surface runoff, the residence time of chloride in surface water, and how variations in weather influence chloride transport and accumulation processes. Monitoring work over three winters showed that the residence time of chloride in small, sewered watersheds varied from 14 to 26 days, depending on winter weather conditions, with 37% to 63% of chloride applied as deicers exported in snowmelt and rainfall surface runoff. In contrast, a monitored highway ditch exported less than 5% of chloride applied to the adjacent road. Stormwater detention ponds were found to act as temporary storage for chloride, with persistent layers of high chloride content at the bottom. Chloride monitoring data and runoff simulations were used to explore the possibility of snowmelt capture as a chloride pollution mitigation strategy. It found that capturing snowmelt runoff close to source areas (roads and parking lots) yields the highest chloride concentrations and removal potential (Herb, Janke, & Herb, 2017).

The impacts of deicing salts—sodium chloride (NaCl), magnesium chloride ($MgCl_2$), and calcium chloride ($CaCl_2$)—on the growth and development of newly hatched rainbow trout (*Oncorhynchus mykiss*) were tested. Rainbow trout were exposed to a wide range of environmentally relevant chloride concentrations (25, 230, 860, 1,500, and 3,000 mg $Cl^- L^{-1}$) over an ecologically relevant time period (25 days). The deicing salts studied were found to have distinct effects. $MgCl_2$ did not affect rainbow trout growth at any concentration. NaCl had no effects at the lowest three concentrations, but rainbow trout length was reduced by 9% and mass by 27% at 3,000 mg $Cl^- L^{-1}$. $CaCl_2$ affected rainbow trout growth at 860 mg $Cl^- L^{-1}$ (5% reduced length; 16% reduced mass) and these effects became larger at higher concentrations (>860 mg $Cl^- L^{-1}$) (11% reduced length; 31% reduced mass). None of the deicing salts affected rainbow trout development. At sub-lethal and environmentally relevant concentrations, the results do not support the paradigm that $MgCl_2$ is the most toxic deicing salt to fish, perhaps due to hydration effects on the Mg^{2+} cation. The results do suggest different pathways for lethal and sub-lethal effects of road salts. Scaled to the population level, the reduced growth caused by NaCl and $CaCl_2$ at critical early-life stages has the potential to negatively affect salmonid recruitment and population dynamics. The findings have implications for environmental policy and management strategies that aim to reduce the impacts of salinization on freshwater organisms (Coldsnow & Relyea, 2018). While these salts have proven effective in melting ice and snow, they can also have negative impacts on the environment, particularly on surface and groundwater quality.

When road salts are applied to roads and highways, they can be carried by runoff into nearby surface water bodies, such as streams, lakes, and rivers. Once in the water, these salts can dissolve and increase the concentration of sodium and chloride ions, which can have a negative impact on aquatic life. High levels of sodium can reduce the growth and survival of certain fish species, whereas high levels of chloride can be toxic to some aquatic plants.

In addition to impacting surface water quality, road salts can also leach into the ground and contaminate groundwater. This can occur when the salt-laden runoff infiltrates into the soil and reaches the water table, which is the layer of soil where water fills all the open spaces. The salt can then move with the groundwater, potentially contaminating drinking water wells and other groundwater resources.

The impacts of road salts on water quality can be significant and long lasting. Research has shown that even low levels of chloride can persist in groundwater for years, and that once a water body is contaminated, it can be difficult and expensive to remediate.

2.3 CONVENTIONAL TECHNIQUES USED FOR TREATMENT AND MANAGEMENT OF RUNOFF WATER FROM DEICING SALT

2.3.1 Treatment and Management

When snow and ice melt, the salt is dissolved and carried by runoff water into storm drains, which eventually discharge into surface water bodies or infiltrate into the ground, potentially contaminating groundwater.

To mitigate the environmental impacts of road salts, various approaches have been developed to treat and manage the runoff water. One common approach is to direct the runoff water to the sewer system for treatment in wastewater treatment plants (WWTPs). This approach is effective in removing dissolved salts from the water, but can also pose challenges to the operation and maintenance of the WWTPs.

When runoff water is directed to the sewer system, it undergoes primary treatment, which involves physical separation of solids and oil using screens, grit chambers, and oil interceptors. The treated water then enters the secondary treatment process, where biological processes are used to remove organic matter, nitrogen, and phosphorus. During the secondary treatment, the dissolved salts in the runoff water are also removed through a process called ion exchange. This process involves passing the water through a bed of resin beads, which exchange the sodium and chloride ions in the water with other ions, such as calcium and magnesium, which are then removed by settling or filtration.

Although directing runoff water to the sewer system and WWTPs is effective in removing dissolved salts from the water, there are some challenges associated with this approach. First, the high salt concentrations in the runoff water can disrupt the biological processes used in the WWTPs, reducing their efficiency and requiring additional energy and resources for treatment. Second, the high salt concentrations can also increase the salinity of the wastewater effluent, which can have negative impacts on the environment if discharged into surface water bodies (Herb, Janke, & Herb, 2017; Szklarek, Górecka, & Wojtal-Frankiewicz, 2022).

To address these challenges, some WWTPs have developed specialized treatment processes to handle the high salt concentrations in the runoff water. For example, some WWTPs use reverse osmosis or electrodialysis to remove dissolved salts from the water before it enters the biological treatment process. These processes involve passing the water through a semi-permeable membrane, which separates the dissolved salts from the water based on their size and charge. This approach can be effective in removing salts from the water, but can also be energy-intensive and costly.

Another approach to managing runoff water from deicing salts is to treat it in specialized purification plants. These plants are specifically designed to handle high salt concentrations and remove other contaminants, such as heavy metals and hydrocarbons. The treatment processes in these plants typically involve chemical precipitation, sedimentation, and filtration, followed by reverse osmosis or electrodialysis to remove dissolved salts from the water. While these plants can be effective in treating runoff water from deicing salts, they can also be costly to operate and maintain (Hintz, Fay, & Relyea, 2022).

In addition to treatment, there are also management approaches that can be used to reduce the environmental impacts of road salts. For example, some municipalities have implemented salt reduction programs, which involve reducing the amount of salt used for deicing by using alternative materials, such as sand or beet juice. Others have implemented better application practices, such as pre-wetting the salt to improve its effectiveness and reduce the amount of salt needed.

While the use of deicing road salts is necessary for winter safety, road salt can also have negative impacts on the environment, particularly on surface and groundwater quality. To mitigate these impacts, various approaches have been developed to treat and manage runoff water from deicing salts, including directing it to the sewer system for treatment in WWTPs and treating it in specialized purification plants.

One of the chloride treatment and management strategies (Herb, Janke, & Herb, 2017) focuses primarily on snowmelt capture, with the idea that capturing small amounts of snowmelt runoff with high chloride concentrations may be a relatively efficient method to mitigate chloride spreading from deicers in the environment (Herb, Janke, & Herb, 2017).

Chloride removal by diversion of saline runoff is most effective (in terms of mass of chloride removed per volume of water) if implemented at the scale of a roadway, before runoff enters the drainage network (e.g. at County Road B). This is illustrated by the larger volumes of water required to remove a similar amount of chloride at the outlet of the Alameda watershed compared to the roadway site (County Road B). Similarly, withdrawal of saline water from the bottom of a pond was found to be less effective than removal at street- or even watershed-level due to dilution of inflow once it enters the pond. Seasonal diversion may also be more effective, and potentially simpler to implement, than the concentration-based diversions. Capturing chloride in snowmelt runoff, by itself, would be relatively ineffective for roadways with pervious ditches, since most of the snowmelt water volume and the chloride it contains is infiltrated. Capturing chloride in runoff from roadways with pervious ditches may require a system of underdrains (Herb, Janke, & Herb, 2017).

Deicing salt loads were primarily born by the surface water. Around 80% of the storm sewer deicing salt discharge (and the majority of highway salt applied) remained in the stream. Stream chloride exceeded the national (UK) Environmental Quality Standard (EQS) for 18% to 33% of the severe winters and peaked at 15% of seawater salinity. Although winter salt loads are often presumed to rapidly flush from catchments and become increasingly dilute further downstream, the (transient) presence of above chloride EQS water may still prove significant and merit study

(Cañedo-Argüelles et al., 2013). Sensitive settings may include (combinations of) streams/rivers of high ecological value, smaller receiving streams characterized by low flows and dilution capacity, low-velocity infiltrating streambed or hyporheic zone exchange waters, and stream water entry and persistence within connected wetland/lake ecosystems. Finally, the study is relevant to the increased adoption of sustainable urban drainage systems (SUDS) and other highway drainage-to-ground schemes that have become increasingly common. The lack of attenuation of conservative chloride poses concern, and schemes should carefully predict the long-term increased salt loads to groundwater resources, especially as problems are slow to manifest and are not easily rectified. Deliberate adjustment of the relative surface water–groundwater loadings of deicing salts demands careful, integrated, and interdisciplinary consideration (Herb, Janke, & Herb, 2017).

Water-sensitive urban design and similar concepts often recommend that a 'treatment train' be employed to improve stormwater quality. The capability of a combined permeable pavement and bioretention basin was examined with a view to developing a permeable pavement reservoir that can supplement the irrigation needs of a bioretention system in semi-arid climates. Salinity was a key study parameter due to published data on salinity in permeable pavement storage, and the potential to harvest water contaminated with deicing salts. To conduct experiments, roof water was collected from a roof in Adelaide, South Australia. Water was amended with NaCl to produce a control runoff (no added salt), a medium (500 mg/L) and a high (1,500 mg/L) salinity runoff. Water was then run through the pavement into the storage reservoir and used to irrigate the bioretention system. Samples were collected from the roof, the pavement reservoir, and the bioretention system outflow to determine whether significant water quality impacts occurred. Results show that while salinity levels increased significantly as water passed through the pavement and through the bioretention system, the increase was beneficial for irrigation purposes as it was from Ca and Mg ions, thus reducing the sodium absorption ratio to levels considered 'good' for irrigation in accordance with several guidelines. Permeable paving increased the pH of water, and this effect was prominent when the initial salt concentration increased. The study shows that permeable pavements with underlying storage can be used to provide supplementary irrigation for bioretention systems, but high initial salt concentrations may present constraints on beneficial use of stormwater (Kazemi, Golzarian, & Myers, 2018).

The seasonal variations of some selected heavy metals (Cd, Cu, Pb, and Zn) and principal anions in soil solutions were monitored as a function of distance from the road at two field sites in Sweden. During the winter, the conductivity and concentrations of dissolved sodium and chloride increased dramatically due to the application of deicing agents (i.e. NaCl). Due to ion exchange, the pH decreased one unit in the soil solutions, whereas the concentrations of total organic carbon decreased due to coagulation and/or sorption to stationary solids. The heavy metal concentrations increased during the winter, but through different mechanisms. Cadmium concentrations in the aqueous phase increased as a response to ion exchange, possibly also enhanced by the formation of chloride complexes. Similarly, the concentrations of zinc increased, due to ion exchange, with calcium and protons. The mechanisms of mobilization for copper and lead were not that clear, probably due to association with coagulated or sorbed organic matter in combination with colloid dispersion (Bäckström et al., 2004). Analysis of tunnel runoff chemical composition revealed two significant trends during the observation period. First, the runoff was dominated by sodium (assumed to be generated from deicing salts), with a pronounced pulse. Lead followed the sodium pulse initially but dropped off quickly. Second, a release of calcium coincided with a drop in the sodium pulse. Release of zinc, copper, and nickel followed the calcium pulse. Results indicated that a small portion of the deposited lead was easily mobilized, whereas the other heavy metals began leaching after the sodium pool (the most responsive component) was exhausted. High levels of zinc (up to 7.5 mg/L) were observed. The levels of other heavy metals measured in this study did not exceed Norwegian drinking-water standards (Andersen & Vethe, 1994).

Winter months in the area of the study site see high usage of deicing salts on roadways and in parking lots (Pineau et al., 2021). February, in particular, is a cold month with high salt application, which then tapers off in March and April. In general, sodium ions may temporarily adsorb to soil particles, but have been shown to wash out easily of soils during infiltration events (Denich et al., 2013). As a result of the high mobility of these ions, bioretention cells are not considered suitable for the removal of deicing salts (Khan et al., 2012). Concentrations of stormwater contaminants revealed that for the event on February 19, 2018, the sodium load was high and minimal release was measured from the bioretention cells. The retention pond, however, was efficient as a buffer. Sodium releases were observed for the March 2018 and April 2018 events, which was expected, given the mobility of this ion, and it was noted that the amount of release was proportional to the amount of rainfall measured on those particular days. Overall, the treatment chain performed better than the bioretention cells on their own, showing a significant decrease in the amount of sodium released during the winter sampling period. However, other studies have indicated that stormwater ponds do not protect surface waters from road salt contamination and suggest that sodium can be released throughout the year (Snodgrass et al., 2017), and that high-intensity rainfall has a strong influence on the amount of salt released from ponds (Barbier et al., 2018; Pineau et al., 2021).

Decentralized treatment of runoff from deicing salts is becoming more common in rural and small communities in cold climates. These communities often have limited resources to invest in large-scale wastewater treatment plants and may need to treat the runoff from deicing salts locally.

One approach to decentralized treatment is the use of constructed wetlands. Wetlands can be designed to remove pollutants from runoff water, including salts and other contaminants. Constructed wetlands consist of shallow basins filled with soil, sand, and gravel, planted with wetland vegetation. As runoff water flows through the wetland, it is filtered by the soil and vegetation, which remove pollutants and nutrients. The filtered water can then be discharged to the environment or used for non-potable purposes.

Another approach to decentralized treatment is the use of stormwater detention ponds. These ponds are designed to hold runoff water from deicing salts and allow it to settle and slowly infiltrate into the ground. The detention ponds can also be designed to promote biological treatment by adding bacteria or other organisms that can break down contaminants in the water.

In addition to constructed wetlands and stormwater detention ponds, there are other decentralized treatment options for runoff water from deicing salts. These include sand filters, bioretention cells, and permeable pavements. Each of these treatment methods has its own advantages and disadvantages and may be more or less suitable depending on the specific site conditions and treatment needs (Hintz, Fay, & Relyea, 2022; Szklarek, Górecka, & Wojtal-Frankiewicz, 2022).

One of the challenges of decentralized treatment of runoff from deicing salts is the high salt content of the water. Salts can be corrosive to infrastructure and equipment and can also have negative impacts on the environment if discharged in high concentrations. To address this challenge, treatment systems must be designed to effectively remove salts from the water. Some treatment methods, such as constructed wetlands, can effectively remove salts, while others, such as sand filters, may require additional treatment steps, such as reverse osmosis or electrodialysis, to remove dissolved salts (Szklarek, Górecka, & Wojtal-Frankiewicz, 2022).

Overall, decentralized treatment of runoff from deicing salts is an important strategy for managing water quality in rural and small communities in cold climates. Although there are challenges to designing and operating effective treatment systems, there are a variety of treatment options available that can be tailored to specific site conditions and treatment needs (Hintz, Fay, & Relyea, 2022).

2.3.2 Alternative Deicing Products

Road salting has major environmental consequences; therefore, it is required to search for an alternative to the road salt. The proposed solution must be cost-effective, leading to less road repairs (Fay et al., 2012). The solution to the road icing problem needs to be mitigated while avoiding the potential negative effects of salting.

The use of the present invention, i.e. AFPs, can solve numerous issues caused by conventional road salting. The use of Tenebrio molitor AFPs in *E. coli* helps to shield the roads from further damage. The AFPs are less dangerous to natural biodiversity. The *E. coli* will be used as a bio-spray to reduce the freezing temperature of water to prevent ice from sticking to the roads during winter. Manning, Pfefferkorn, and Raviraja (2019) proposes the addition of Origami 2 to solve the issue at hand. Creating a bio-spray will prove to be an effective and eco-friendly solution. The solution presented in this paper also has other benefits such as accident risk-reduction on the roads and less damage to the roads compared to road salts.

Several representative deicing salts were investigated (Ke, Zhang, & Tian, 2019), and the effects of deicing ability, salt freezing corrosion on concrete, and plant growth were comprehensively tested. Finally, the decision of deicing salt was made based on an analytic hierarchy process (AHP). Results showed that deicing salts achieving the best deicing effect were not the same at different concentrations. Deicing salts of 3% concentration have the greatest corrosion to concrete. Magnesium chloride and calcium magnesium acetate have the least impact on plants among all of the studied deicing salts. Using AHP, ethylene glycol and calcium magnesium acetate are selected as optimal items under different priorities (Ke, Zhang, & Tian, 2019).

Freshwater organisms with heightened sensitivity to salt would be expected to benefit from the reduction in road salt use triggered by a transition to alternative deicing products, but to determine if the potential for harm is reduced, there is a need to assess whether the replacement products themselves pose a risk to aquatic biota. The objective of this study (Gillis et al., 2021) was to determine the relative toxicity of three road salt alternatives, including salt brine, a beet juice product, and a brine–beet juice product, to early life stage freshwater mussels. The observed toxicities are examined in relation to the various constituents of each product and their application rates for winter road maintenance.

The negative impact of road salt on freshwater ecosystems has led to an interest in “eco-friendly” deicing products. Biota, including freshwater mussels that have heightened salt sensitivity, would be expected to benefit from a transition to alternative deicing products. However, it was unknown whether the alternatives themselves pose a risk. The toxicity of three road salt alternatives, including a salt brine, a beet juice, and a brine–beet juice product were examined. Lampsilis fasciola glochidia (larvae) were exposed to dilutions (0–2%) of deicing products. On a per volume basis, beet juice products were significantly more toxic than brine, with 48-hour EC50s (95% confidence intervals) as follows: brine, 0.42% (0.35–0.50%); beet juice, 0.020% (0.018–0.022%); and brine–beet juice, 0.034% (0.028–0.039%). Unlike brine, beet juice toxicity did not correspond with the concentration of chloride in the exposure. While elevated trace metals (Cu, Fe, Zn) and reduced water quality occurred in the 1% and 2% beet juice exposures, toxicity occurred at much lower dilutions ($\leq$0.05%). The toxicity of beet juice products aligned with glochidia potassium EC50s. Based on toxicity and application rates, beet juice deicing products pose more of a hazard to early life stage mussels than traditional products and could contribute substantial potassium to receiving environments (Gillis et al., 2021).

Current regulations are generally inadequate because they are regionally inconsistent, lack legal consequences, and have few ion-specific standards. The lack of ion-specific standards is problematic, because each anthropogenic source of freshwater salinization is associated with a distinct set of ions that can present unique social and economic costs. Additionally, the environmental and toxicological consequences of freshwater salinization are often dependent on the occurrence, concentration, and ratios of specific ions. Therefore, to protect fresh waters from continued salinization, discrete, ion-specific management and regulatory strategies should be considered for each source of freshwater salinization, using data from standardized, ion-specific monitoring practices. To develop comprehensive monitoring, regulatory, and management guidelines, we recommend the use of co-adaptive, multi-stakeholder approaches that balance environmental, social, and economic costs and benefits associated with freshwater salinization (Schuler et al., 2019).

2.4 RECOVERY AND REUSE OF DEICING SALT

The recovery and reuse of deicing salt from runoff water can provide significant benefits in terms of cost savings, reduced environmental impact, and improved sustainability. However, there are also challenges associated with the implementation of these technologies. One of the main challenges is the cost of the recovery and reuse systems, which can be significant. This may be a barrier to the widespread adoption of these systems, particularly in areas with limited resources. The initial investment required to implement recovery and reuse systems may be higher than the cost of purchasing new salt. Additionally, there may be ongoing maintenance and operational costs associated with these systems that need to be considered.

One approach to the recovery and reuse of deicing salt from runoff water is the use of constructed wetlands. Constructed wetlands are shallow, vegetated water bodies designed to treat wastewater through natural processes such as filtration, sedimentation, and microbial degradation. In the case of deicing salt recovery, constructed wetlands can be designed to promote the growth of salt-tolerant plants that can accumulate and store the salt from the runoff water. The harvested plant material can then be used as a source of deicing salt. Constructed wetlands have been shown to be effective in the recovery of deicing salt, with up to 100% of the salt being recovered. However, the process requires significant land area and is not suitable for all sites.

Another challenge is the potential for contamination of the recovered salt. The recovered salt may contain impurities or other materials that could affect its effectiveness as a deicer. It is important to ensure that the recovered salt meets certain standards before it is reused. This may require additional testing and quality control measures, which can add to the overall cost of the system.

Furthermore, the effectiveness of the recovery and reuse systems may be affected by weather conditions. For example, if there is a prolonged period of sub-zero temperatures, the runoff water may freeze, making it difficult to recover the salt. Similarly, if there is heavy snowfall or rainfall, the concentration of salt in the runoff water may be diluted, reducing the amount of salt that can be recovered.

Despite these challenges, there are several technologies and methods available for the recovery and reuse of deicing salt. One common method is the use of settling basins, which allow the salt to settle to the bottom of the basin while the water is drained off and treated separately. The recovered salt can then be washed and dried before being reused.

Another method is the use of ion exchange resins, which can be used to selectively remove salt ions from the runoff water. The recovered salt can then be washed and dried before being reused. This method has the advantage of producing a high-quality salt product, but it may be more expensive than settling basins.

Membrane filtration is another technology that can be used to recover deicing salt from runoff water. This method involves passing the runoff water through a membrane that selectively removes the salt ions. The recovered salt can then be washed and dried before being reused. This method has the advantage of being able to recover a higher percentage of the salt from the runoff water, but it may be more expensive than settling basins or ion exchange resins.

In addition to these technologies, there are also several strategies that can be used to improve the effectiveness and efficiency of recovery and reuse systems. One strategy is to improve the design of the road and drainage systems to minimize the amount of runoff water that is generated. This can reduce the amount of salt that needs to be recovered and reused, as well as reduce the overall cost of the system. Another strategy is to use brine solutions instead of solid salt for deicing. Brine solutions can be applied directly to the road surface, reducing the amount of salt that is needed and minimizing the amount of runoff water that is generated. This can make it easier and more cost-effective to recover and reuse the salt.

Overall, the recovery and reuse of deicing salt from runoff water can provide significant benefits in terms of cost savings, reduced environmental impact, and improved sustainability. While there are challenges associated with the implementation of these systems, there are also several technologies and strategies available to address these challenges. By carefully considering the costs and benefits of recovery and reuse systems, transportation agencies and local governments can make informed decisions about the most appropriate approach for their specific needs and circumstances.

2.5 CONCLUSIONS AND PERSPECTIVES

Ensuring road safety in cold climates through deicing salt usage is vital, but its untreated runoff water can pose significant environmental risks. Decentralized treatment systems offer a cost-effective and practical solution that can be implemented in diverse locations to mitigate these environmental impacts. Technologies such as constructed wetlands, permeable pavements, bioretention systems, and vegetated infiltration systems have demonstrated their ability to effectively remove pollutants from runoff water.

To achieve optimal performance, the design and operation of decentralized treatment systems must consider various factors, including the types and concentrations of pollutants, hydraulic loading rates, temperature fluctuations, and maintenance requirements. These considerations will guarantee the efficient removal of pollutants and minimize the environmental harm caused by deicing salt runoff. The advantages of decentralized systems are manifold, including reduced infrastructure needs, cost-effectiveness, and the capacity to target specific pollutants, addressing environmental concerns more precisely.

However, the success of decentralized treatment systems relies heavily on effective maintenance and monitoring. Challenges may arise due to limited manpower or expertise in some areas, making it imperative to develop practical and user-friendly maintenance protocols. Regular monitoring and evaluation of these systems are also essential to assess their performance and identify any areas for improvement.

In addition to decentralized treatment systems, the reduction of deicing salt usage on roads presents an important opportunity to mitigate environmental impacts further. By exploring alternative materials like beet juice or calcium magnesium acetate, the harmful effects of traditional deicing salts can be minimized. Furthermore, optimizing the application of deicing salt through improved timing and amount can reduce overall usage while maintaining road safety standards. Combining these efforts with decentralized treatment systems can create a comprehensive approach to address environmental concerns associated with deicing salt runoff.

The development and implementation of decentralized treatment systems are crucial steps toward improving the environmental sustainability of road deicing practices. Treating runoff water at the source can significantly reduce the volume of contaminated water entering waterways, thus minimizing downstream contamination and safeguarding aquatic ecosystems. By adopting decentralized treatment solutions, we can proactively address the negative impacts of deicing salt on the environment and human health.

In conclusion, decentralized treatment systems offer a practical and efficient means to mitigate the environmental damage caused by deicing salt runoff. Their diverse range of technologies provides tailored solutions to address specific pollutants effectively. While challenges related to maintenance and monitoring must be tackled, the overall advantages of these systems outweigh the drawbacks. Coupled with the reduction of deicing salt usage and improvements in application techniques, decentralized treatment approaches can usher in a new era of sustainable road deicing practices. By investing in further research, development, and widespread implementation of these technologies, we can safeguard our environment and promote the well-being of both ecosystems and communities in cold climate regions.

REFERENCES

Andersen, S., and Ø Vethe. 1994. "Mobilisation of heavy metals during tunnel maintenance." *Science of The Total Environment* 146-147:479–483. https://doi.org/10.1016/0048-9697(94)90272-0.

Ashforth, D., and N. D. Yan. 2008. The interactive and effects of calcium concentration and temperature on the survival of Daphnia pulex at high and low food concentrations reproduction. *Limnol. Oceanogr.* 53:420–432. doi:10.4319/lo.2008.53.2.0420

Bäckström, M., S. Karlsson, L. Bäckman, L. Folkeson, and B. Lind. 2004. "Mobilisation of heavy metals by deicing salts in a roadside environment." *Water Res* 38 (3):720–732. doi: 10.1016/j.watres.2003.11.006.

Barbier L, Suaire R, Durickovic I, Laurent J and Simonnot MO (2018) Is a road stormwater retention pond able to intercept de-icing salt? Water, Air and Soil Pollution 229(251), https://doi.org/10.1007/s11270-018-3908-9.

Bridgeman, T. B., Wallace, C. D., Carter, G. S., Carvajal, R., Schiesari, L. C., Aslam, S., ... Kling, G. W. (2000). A limnological survey of Third Sister Lake Michigan with historical comparisons. *Journal of Lake and Reservoir Management*, 16: 253–267. https://doi.org/10.1080/07438140009354234

Cañedo-Argüelles, M., Bundschuh, M., Gutierrez-Canovas, C., Kefford,B. J., Prat, N., Trobajo, R., & Schaefer, R. B. (2014). Effects of repeated salt pulses on ecosystem structure and functions in a stream me-socosm. *Science of the Total Environment*, 476: 634–642. https://doi.org/10.1016/j.scitotenv.2013.12.067

Cañedo-Argüelles, M., Kefford, B. J., Piscart, C., Prat, N., Schaefer, R. B., & Schulz, C.-J. (2013). Salinisation of rivers: An urgent ecological issue. *Environmental Pollution*, 173: 157–167. https://doi.org/10.1016/j.envpol.2012.10.011

Cañedo-Argüelles, M., Sala, M., Peixoto, G., Prat, N., Faria, M., Soares, A. M. V. M., Kefford, B. (2016). Can salinity trigger cascade effects on streams? A mesocosm approach. *Science of the Total Environment*, 540: 3–10. https://doi.org/10.1016/j.scitotenv.2015.03.039

Coldsnow, Kayla D, and Rick A Relyea. 2018. "Toxicity of various road-deicing salts to Asian clams (*Corbicula fluminea*)." *Environmental Toxicology and Chemistry* 37 (7):1839–1845.

Denich C, Bradford A and Drake J (2013) Bioretention: assessing effects of winter salt and aggreagte application on plant health, media clogging and effluent quality. *Water Quality Research Journal of Canada* 48(4):387–399, https://doi.org/10.2166/wqrjc.2013.065.

Dickmann, M. D., & Gochnauer, M. B. (1978). Impact of sodium chloride on the microbiota of a small stream. *Environmental Pollution*, 17:109–126. https://doi.org/10.1016/0013-9327(78)90044-7

Evans, Marlene, and C. M Frick. 2001. "The effects of road salts on aquatic ecosystems." Environment Canada, Water Science and Technology Directorate, Ottawa, ON, Canada

Fay, Laura, Michelle Akin, and Xianming Shi. 2012. *Cost-Effective and Sustainable Road Slope Stabilization and Erosion Control*. Editor: Transportation Research Board Publisher. National Academies Press. DOI:10.13140/2.1.1229.1525

Findlay, Stuart EG, and Victoria R Kelly. 2011. "Emerging indirect and long-term road salt effects on ecosystems." *Annals of the New York Academy of Sciences* 1223 (1):58–68.

Gillis, Patricia Leigh, Joseph Salerno, C. James Bennett, Yaryna Kudla, and Margot Smith. 2021. "The relative toxicity of road salt alternatives to freshwater mussels; Examining the potential risk of eco-Friendly de-icing products to sensitive aquatic species." *ACS EST Water* 1 (7):1628–1636. doi: 10.1021/acsestwater.1c00096.

Herb, William, Ben Janke, and William Herb. 2017. "Study of De-icing Salt Accumulation and Transport Through a Watershed." Research Project. Final Report 2017-50. Sponsored by: Minnesota Local Road Research Board, USA http://www.dot.state.mn.us/research/reports/2017/201750.pdf

Hessen, D. O., N. E. W. Alstad, and L. Skardal. 2000. Calciumlimitation in Daphnia magna. *J. Plankton Res.* 22: 553–568. doi:10.1093/plankt/22.3.553

Hintz, William D, Laura Fay, and Rick A Relyea. 2022. "Road salts, human safety, and the rising salinity of our fresh waters." *Frontiers in Ecology and the Environment* 20 (1):22–30.

Hintz, William D, and Rick A Relyea. 2019. "A review of the species, community, and ecosystem impacts of road salt salinisation in fresh waters." *Freshwater Biology* 64 (6):1081–1097.

Hocking, RK, RSB Greene, CE Hughes, SG Johnston, and S Grover. 2011. "The impact of de-icing salts on alpine bogs in Kosciuszko National Park." Paper presented to the 2011 International Union of Geodesy and Geophysics (IUGG) General Assembly, Earth on the Edge: Science for a Sustainable Planet, 28 June–7 July 2011, Melbourne, Australia. Retrieved from http://iugg.org/archive/iugg2011/tourhosts.com.au/live/iugg2011/pdf/IUGG2011_FinalEntireProgram.6June.pdf

Jensen, Thomas, Sondre Meland, Ann Schartau, and Bjørn Walseng. 2014. "Does road salting confound the recovery of the microcrustacean community in an acidified lake?" *Science of The Total Environment* 478:36–47. doi: 10.1016/j.scitotenv.2014.01.076.

Judd KE, Adams HE, Bosch NS et al. (2005) A case history: Effects of mixing regime on nutrient dynamics and community structure in Third Sister Lake, Michigan during late winter and early spring 2003. *Lake Reserv Manag* 21:316–329.

Kazemi, Fatemeh, Mahmood Reza Golzarian, and Baden Myers. 2018. "Potential of combined water sensitive urban design systems for salinity treatment in urban environments." *Journal of Environmental Management* 209:169–175. https://doi.org/10.1016/j.jenvman.2017.12.046.

Khan UT, Valeo C, Chu A and Duin BV (2012) Bioretention cell efficacy in cold climates: Part 2 – water quality performance. *Canadian Journal of Civil Engineering* 39(11): 1222–1233, https://doi.org/10.1139/l2012-111.

Ke, Guoju, Jun Zhang, and Bo Tian. 2019. "Evaluation and selection of de-icing salt based on multi-factor." *Materials* 12 (6):912.

Ledford, S. H., Lautz, L. K., & Stella, J. C. (2016). Hydrogeologic processes impacting storage, fate, and transport of chloride from road salt in urban riparian aquifers. *Environmental Science & Technology* 50:4979–4988. https://doi.org/10.1021/acs.est.6b00402

Manning, Lola, Eva Pfefferkorn, and Samiksha Raviraja. 2019. "Using Antifreeze Proteins to Develop an Alternative to Road Salts." BioBuilder Club, Renaissance School, Charlottesville, Virginia, USA. BioTreks | www.biotreks.org. 4(1) e201906. http://biotreks.org/wp-content/uploads/2019/10/e201906.pdf

Mason, C. F., Norton, S. A., Fernandez, I. J., & Katz, L. E. (1999). Deconstruction of the chemical effects of road salt on stream water chemistry. *Journal of Environmental Quality* 28:82–91. https://doi.org/10.2134/jeq1999.00472425002800010009x

Pineau, Béatrice, Caroline Brodeur-Doucet, Juliette Corrivault-Gascon, Diane Arjoon, Paul Lessard, Geneviève Pelletier, and Sophie Duchesne. 2021. "Performance of green infrastructure for storm water treatment in cold climate (Canada)." *Journal of Environmental Engineering and Science* 16 (4):185–194.

Porter-Goff, E. R., Frost, P. C., and Xenopoulos, M. A. (2013). Changes in riverine benthic diatom community structure along a chloride gradient. *Ecological Indicators*, 32:97–106. https://doi.org/10.1016/j.ecolind.2013.03.017

Richburg, J. A., Patterson, W. A., and Lowenstein, F. (2001). Effects of road salt and *Phragmites australis* invasion on the vegetation of a western Massachusetts calcareous lake-basin fen. *Wetlands* 21:247–255. https://doi.org/10.1672/0277-5212(2001)021[0247:eorsap]2.0.co;2

Schuler, Matthew S, Miguel Cañedo-Argüelles, William D Hintz, Brenda Dyack, Sebastian Birk, and Rick A Relyea. 2019. "Regulations are needed to protect freshwater ecosystems from salinization." *Philosophical Transactions of the Royal Society B* 374 (1764):20180019.

Shenton, Mark D, Susan J Nichols, Jon P Bray, Benjamin JG Moulding, and Ben J Kefford. 2022. "The effects of road de-icing salts on water quality and macroinvertebrates in Australian alpine areas." *Archives of Environmental Contamination and Toxicology* 82 (2):266–280.

Snodgrass JW, Moore J, Lev SM et al. (2017) Influence of modern stormwater management practices on transport of road salt to surface waters. *Environmental Science & Technology* 51(8):4165–4172, https://doi.org/10.1021/acs.est.6b03107

Soleimanifar, Parisa. 2019. "Effects of Road Salt Applications on Human and Ecological Health: A Risk Assessment and Risk Management for Prospective Policy Options." Presented to the Advanced Academic Program Environmental Policy and Climate Capstone. Johns Hopkins Krieger School of Arts and Sciences. https://jscholarship.library.jhu.edu/server/api/core/bitstreams/e56bb987-560a-439d-bd9d-9a2d301875ad/content

Stefan, Heinz G, Eric Novotny, Andrew Sander, and Omid Mohseni. 2008. "Study of environmental effects of de-icing salt on water quality in the twin cities metropolitan area, Minnesota."

Szklarek, Sebastian, Aleksandra Górecka, and Adrianna Wojtal-Frankiewicz. 2022. "The effects of road salt on freshwater ecosystems and solutions for mitigating chloride pollution—a review." *Science of The Total Environment* 805:150289. https://doi.org/10.1016/j.scitotenv.2021.150289.

Tromp, Karin, Ana T. Lima, Arjan Barendregt, and Jos T. A. Verhoeven. 2012. "Retention of heavy metals and poly-aromatic hydrocarbons from road water in a constructed wetland and the effect of de-icing." *Journal of Hazardous Materials* 203-204:290–298. https://doi.org/10.1016/j.jhazmat.2011.12.024.

Valleau, Robin E., Andrew M. Paterson, and John P. Smol. 2020. "Effects of road-salt application on Cladocera assemblages in shallow Precambrian Shield lakes in south-central Ontario, Canada." *Freshwater Science* 39 (4):824–836. https://doi.org/10.1086/711666.

Van Meter, R. J., Swan, C. M., and Trossen, C. A. (2012). Effects of road deicer (NaCl) and amphibian grazers on detritus processing in pond mesocosms. *Environmental Toxicology and Chemistry* 31:2306–2310. https://doi.org/10.1002/etc.1949

Vítězová, Monika, Nuttakan Nitayapat, and Tomáš Vítěz. 2018. "Temperature and de-icing salt, effect on the activated sludge respiration." *CLEAN—Soil, Air, Water* 46 (11):1800050. https://doi.org/10.1002/clen.201800050

Wallace, A. M., and Biastoch, R. G. (2016). Detecting changes in the benthic invertebrate community in response to increasing chloride in streams in Toronto, Canada. *Freshwater Science* 35:353–363. https://doi.org/10.1086/685297

3 Mobile and Portable Technologies for Water/Wastewater Treatment in Cold Environments

Song Yan[a], *Patrick Drogui*[a], *Rajeshwar D. Tyagi*[b,c], *and Jonathan W.C. Wong*[b]
[a] Université du Québec, Québec, Canada
[b] Dongguan University of Technology, Dongguan, China
[c] BOSK Bioproducts, Québec, Canada

3.1 INTRODUCTION

Mobile and portable water and wastewater treatment systems are designed to provide emergency or temporary solutions for situations such as construction sites or long-term water treatment requirements. These systems can be installed and operated independently or combined to form complete treatment plants with advanced technologies for water and wastewater treatment and purification.

These mobile systems utilize various technologies, including petrochemical, ion exchange, membrane, multimedia filtration, and membrane bioreactors (MBR). The systems are usually housed in trailers, making them portable and easy to transport to different locations as needed (Park et al. 2015).

Ensuring effective wastewater treatment is essential for safeguarding both the environment and human health. There are two primary methods of removing wastewater contaminants: mechanical processes and passive processes that rely on natural systems, such as wetlands or marine waters. However, in colder climates, removing wastewater contaminants can pose a significant challenge. The extreme weather conditions often preclude the use of many systems commonly employed in warmer climates. Moreover, in remote areas, where road access is limited, costs increase and timely maintenance, repairs, and monitoring become even more difficult (Daley et al. 2015).

Designing, constructing, and operating wastewater collection systems in cold climate conditions can be a significant challenge. The extreme cold can cause fluids to freeze, requiring heating equipment to maintain them above their freezing points, and electrical equipment can suffer at temperatures below 5°C. This makes it imperative to consider the use of heating equipment to mitigate these issues. In particular, the efficiency of biological treatment processes can be significantly affected by the cold climate, and chemical processes may also be impacted (Smith 1996; Gunnarsdóttir et al. 2013).

Mobile and portable water and wastewater treatment technologies are designed to be flexible, compact, and easily automated, making them suitable for use in any setting. They can be equipped with various options, and their mobility is convenient for use in construction sites.

The advantages of these systems include their ability to be quickly delivered, their capacity to treat water from various sources, and their ability to adapt to seasonal changes in water and wastewater. However, in cold environments, effective water and wastewater treatment can be challenging due to the extreme temperatures that may cause certain systems, commonly used in normal temperatures, to become unusable (Banas et al. 2005; Northcott et al. 2007; Park et al. 2015).

DOI: 10.1201/9781003376415-4

In terms of operation and maintenance, a mobile water treatment system should be easy to transport quickly to the desired location and require straightforward maintenance and operation procedures (Lakho et al. 2020; Lakho et al. 2021; Lakho et al. 2022).

This chapter presents a comprehensive review of mobile and portable technologies for water and wastewater treatment in cold environments. Several sample processes are highlighted in Table 3.1 to illustrate the different technologies discussed in this chapter. The chapter covers the design, operation, and application of these technologies, emphasizing their treatment performance in physical, chemical, microbiological, and other advanced treatment methods. Case studies are included to highlight the challenges encountered during the processes, and perspectives for future developments are discussed.

TABLE 3.1
Summary of Various Mobile and Portable Technologies for Water/Wastewater Treatment in Cold Environments

SN	Technology	Before Treatment	After Treatment, or Treatment Efficiency	References
			Drinking Water	
1	Ultraviolet water purification systems		A 5.9 log reduction at 10 minutes (which was a total kill) using a power level of P1 and 0.5P1 (equating to 8.73E05 CFU)	(Banas et al. 2005)
2	MF PCF	Turbidity: 13	MF: 0.07 PCF: 0.1	(Park et al. 2015)
3	MF + AC PCF + AC MF + RO PCF + RO	COD (mg/L): 6	(mg/L): MF + AC: 4 PCF + AC: 4.1 MF + RO: 0 PCF + RO: 0	(Park et al. 2015)
4	MF + RO PCF + RO	TN (mg/L): 5.33	(mg/L): MF + RO: 0.22 PCF + RO: 0.27	(Park et al. 2015)
5	MF + RO PCF + RO	NH_3-N (mg/L): 1.71	(mg/L): MF + RO: 0 PCF + RO: 0	(Park et al. 2015)
6	MF + RO PCF + RO	NO_3-N (mg/L): 2.34	(mg/L): MF + RO: 0 PCF + RO: 0	(Park et al. 2015)
7	MF + RO PCF + RO	TP (mg/L): 0.19	(mg/L): MF + RO: 0.12 PCF + RO: 0.12	(Park et al. 2015)
8	MF + RO PCF + RO	DOC (mg/L): 2.11	(mg/L): MF + RO: 0.25 PCF + RO: 0.29	(Park et al. 2015)
9	Ultrafiltration and reverse osmosis (UF-RO) membranes to produce potable water		Reduced the nitrite from 4.8 $mg{\cdot}L^{-1}$ to below 0.03 $mg{\cdot}L^{-1}$, nitrate from 12.5 $mg{\cdot}L^{-1}$ to below 2.0 $mg{\cdot}L^{-1}$, and ammonium from 3.8 $mg{\cdot}L^{-1}$ to below 0.2 $mg{\cdot}L^{-1}$.	(Lakho et al. 2020)
			Wastewater	
10	A mobile vertical flow constructed wetland (MCW)	COD: 43 $mg{\cdot}L^{-1}$ BOD: 16 $mg{\cdot}L^{-1}$ TSS: 2.7 $mg{\cdot}L^{-1}$ TP: 1.7 $mgP{\cdot}L^{-1}$	Removal of 90% (COD), 95% (BOD), 97% (TSS), and 76% (TP), respectively; total nitrogen removal was 25% (from 45 $mgN{\cdot}L^{-1}$ to 34 $mgN{\cdot}L^{-1}$).	(Lakho et al. 2020).
11	Up-flow anaerobic sludge blanket reactors (UASB) reactors (25, 16, 12, 8.5, 5.5, and 2.5°C)		Stable COD removal efficiencies of 50–70% were achieved at 25°C down to 8.5°C. COD removal efficiencies were reduced at temperatures below 8.5°C, but significant methane formation was observed even at 2.5°C.	(Safitri 2022)

(Continued)

TABLE 3.1 (Continued)
Summary of Various Mobile and Portable Technologies for Water/Wastewater Treatment in Cold Environments

SN	Technology	Before Treatment	After Treatment, or Treatment Efficiency	References
12	UASB without and with ($AnMBR_{UASB(UF)}$) membrane		Both systems met the EU chemical oxygen demand (COD) effluent standard (<125.0 mg L^{-1}).	(Petropoulos et al. 2021).
13	Biochar packed reactors		Different nitrogen-loading (20, 30, and 50 mg TN/L) biochar packed reactors (C:N:P = 100:5:1) within 125 days at 5°C. The results showed that high nitrogen loading resulted in an NH_4^+ (TN) removal efficiency decline from 98% (57%) to 83% (29%), effluent dissolved organic nitrogen (DON) concentrations of R20, R30, and R50 were 0.44±0.18, 0.85±0.35, and 0.66±0.26 mg/L, respectively.	(He et al. 2018)
14	Pretreatment of domestic sewage in an anaerobic filter (AF) and an anaerobic hybrid (AH) a two-step system, AF + AH (with granular sludge) reactor		A CODt removal of 71% was achieved with 60% conversion to methane from the removed CODt when the AF + AH system was operated at an HRT of 4 + 8 h at 13°C.	(Elmitwalli, Zeeman, and Lettinga 2001)

MF: microfiltration; NF: nanofiltration; PCF: pore control fiber filter; UV: ultraviolet; AC: activated carbon; RO: reverse osmosis; UF: ultrafiltration; COD: chemical oxygen demand; BOD: biological oxygen demand; TSS: total suspended solids; TP: total phosphorus; TN: total nitrogen; DON: dissolved organic nitrogen; CFU: colony forming unit; DOC: dissolved organic carbon.

3.2 MOBILE AND PORTABLE TECHNOLOGIES FOR DRINKING WATER TREATMENT IN COLD ENVIRONMENTS

3.2.1 Membrane Technology

A mobile water treatment package system comprises an alternative for stable emergency water supply in response to various situations. Such a mobile system was developed (Park et al. 2015) to ensure mobility by simplifying process configuration and realizing modularization of various process technologies. Based on various combinations of unit processes, they were presented according to each treatment process, indicating the possibility of achieving the water quality objectives. Water treatment unit processes were established by setting water quality objectives. Based on the classification of water treatment unit process of pretreatment or primary treatment (filtration using pore control fiber filter (PCF), microfiltration (MF), reverse osmosis (RO)), and post-treatment (filtration process using activated carbon (AC) and disinfection process using ultraviolet (UV)), four water treatment process combinations were proposed: (1) PCF + AC + UV; (2) PCF + RO + UV; (3) MF + AC + UV; (4) MF + RO + UV. Each process combination was made to ensure the processing capacity of 30 m^3/day (Park et al. 2015). In addition, a mobile water treatment package system was manufactured using two 20-ft containers compliant with the International Organization for Standardization (IOS) standards for increased mobility.

The analysis of removal efficiency showed that the MF and PCF were very effective in removing turbidity. Around 99% removal of turbidity was observed by the PCF and MF. For high-turbidity water, the MF was more efficient than the PCF. One of the reasons why MF may be more efficient than PCF for high turbidity water is likely because MF has a smaller pore size than PCF, which allows it to remove smaller particles and impurities that may be present in high turbidity water. In contrast, PCF may not be able to effectively remove all of the larger particles present in high-turbidity water. The specific reasons may depend on the specific application and the characteristics of the water being treated. AC was effective in removing chemical oxygen demand (COD) and total nitrogen (TN), while RO was effective in removing COD, TN, NH_3-N, total phosphorus (TP), and DOC (Park et al. 2015). The unit processes in the mobile water treatment package prototype can be combined in a selective manner, based on the specific properties of the raw water being treated. This allows for automated selection and operation of the unit processes.

It also showed that power consumption by each process can be used for fabricating the power unit (diesel or gasoline fuel module to generate power, photovoltaic module, etc.) for self-generation in the future. The system is judged to be applied for small-scale water supplies as well as for emergency use (Park et al. 2015).

Therefore, in the cold climate conditions, if heating equipment is considered, this type of mobile water treatment package system could be used to generate the water needed (Northcott et al. 2003; Northcott et al. 2007; Kabeel and El-Said 2015; Kobayashi et al. 2020; Zhang et al. 2021).

Temporary events such as music festivals are often held in places that are not connected to a sewage network. A mobile vertical flow constructed wetland (MCW) with an area of 15 m^2 was constructed (Lakho et al. 2020), operated, and optimized for three years treating grey water (GW) as well as grey and black water (GW + BW) at different festival locations to tackle this problem.

As a further development, the MCW was coupled to a drinking water treatment system using ultrafiltration and reverse osmosis (UF-RO) membranes to produce potable water on-site. The drinking water system produced potable water that met the legislative criteria. As such, a sustainable and mobile water treatment system aiming at producing potable water at temporary events was demonstrated (Lakho et al. 2020; Lakho et al. 2021; Lakho et al. 2022). The mobile potable water production unit was installed treating the effluent of the MCW. The aim was to produce potable water by removing all harmful components (e.g. metals and organic micropollutants). Nitrogen components were the main focus due to (1) the insufficient nitrogen removal (about 25%) in the MCW; (2) the fact that nitrite and ammonium are considered as toxic, and (3) because nitrite and ammonium are small molecules that can be used as probe components for RO performance assessment. It showed that the water produced met the standards for potable water as N removal proved quite satisfying. A mixed-bed ion-exchange cartridge after RO was installed, which reduced the nitrite from 4.8 mg/L to below 0.03 mg/L, nitrate from 12.5 mg/L to below 2.0 mg/L, and ammonium from 3.8 mg/L to below 0.2 mg/L (Lakho et al. 2020).

The challenge tests showed that the MCW could be operated at a maximum hydraulic loading rate of 1.1 $m^3 \cdot m^{-2} \cdot d^{-1}$. During the treatment of GW, the MCW achieved effluent concentrations of 43 mg/L, 16 mg/L, 2.7 mg/L, and 1.7 mg/L for COD, biological oxygen demand (BOD), total suspended solids (TSS), and TP, respectively. Removal of COD, BOD, and TSS from grey water was 90%, 95%, and 97%, respectively. This corresponds to a removal of 90% COD, 95% BOD, 97% TSS, and 76% TP, respectively. Total nitrogen removal was 25% (from 45 $mgN \cdot L^{-1}$ to 34 $mgN \cdot L^{-1}$) as particular denitrification was not complete (Lakho et al. 2020). Removal of phosphorus and nitrogen from grey water was 76% and 25%, respectively. The effluent of the constructed wetland was made safe to drink after it went through the treatment process (Lakho et al. 2020).

Lakho et al. (2022) evaluated the life cycle impact of two decentralized water treatment technologies (DWTs) in Belgium, using the Life Cycle Assessment tool. The first technology was a mobile constructed wetland that treated grey water (Scenario 1) at music festivals, producing 100 m^3 of potable water from 400 m^3 of wastewater generated per festival. This was coupled with a membrane-based drinking water production system. The second technology was a vertical flow constructed wetland that treated black water (Scenario 2) at a restaurant with 135 visitors per day, also coupled with a membrane system. The study compared these two technologies to conventional alternatives (PET bottled water and public drinking water supply, respectively) and found that in most impact categories, Scenarios 1 and 2 had significantly lower impact than their conventional alternatives. Sensitivity analysis was also performed, which showed that the DWTs were still environmentally feasible compared to their conventional alternatives at the shortest distance studied (Scenario 1: 175 km and Scenario 2: 75 m). These results indicated that DWTs could be considered environmentally beneficial under certain conditions. The system was also found to be suitable for use in cold climate conditions (Lakho et al. 2022).

3.2.2 Electrochemical Technology

A mobile water treatment system was developed for remote cold regions contaminated sites (Northcott et al. 2003; Northcott et al. 2005; Northcott et al. 2007). Due to the fact that in remote cold regions, the handling, transport, and eventual disposal of material is very difficult and costly.

The mobile water treatment plant (Northcott et al. 2003; Northcott et al. 2005; Northcott et al. 2007) comprised a multistage treatment process for the removal of colloidal particles and suspended solids, via coagulation and settling (electrocoagulation process), and dissolved metal contaminants using ion-exchange columns. To understand how treated water quality is affected, a mechanistic approach was taken that examined various operating parameters. These parameters include chemical dose, water temperature, inlet turbidity, and pH, which all impact the coagulation process and the physicochemical nature of the flocs that are produced (Northcott et al. 2003; Northcott et al. 2005; Northcott et al. 2007).

Flocculation effectiveness is reduced at lower water temperatures due to increased liquid viscosity, slower precipitation kinetics, and smaller particle aggregates that settle less efficiently. Generally, ferric chloride is more effective than alum at low temperatures (Letterman 1999; Northcott et al. 2003; Northcott et al. 2005; Northcott et al. 2007; Faust and Aly 2018). To achieve proper flocculation in waters (from site), a longer detention time of at least 15 minutes was required due to the low temperature. This is necessary to allow for the formation of large, fast-settling flocs. To ensure that the system is efficient when scaled up, the mixing time used in the laboratory tests will be doubled in the full-scale water treatment system (Northcott et al. 2003; Northcott et al. 2005; Northcott et al. 2007).

The laboratory and full-scale water treatment tests (Northcott et al. 2003; Northcott et al. 2005; Northcott et al. 2007) found that, in cold regions with contaminated water, the temperature of coagulation did not significantly affect the formation of flocs. However, the increased viscosity of water at low temperatures requires careful consideration of the flow rate through the settler to ensure adequate particle settling. In comparison, the presence of seawater in suspension has a more significant impact on coagulation than low temperature. This finding is particularly relevant for contaminated sites located in coastal areas, which may be periodically exposed to marine inundation (Northcott et al. 2003; Northcott et al. 2005; Northcott et al. 2007).

Ambauen et al. (2020) conducted a study to investigate the feasibility of using electrochemical oxidation (EO) for removing organic compounds from water in cold climate regions. To determine the optimal conditions for EO treatment, a two-level full factorial design was developed. The study focused on examining the effects of three factors, namely, temperature (6/20°C), pH (7/10), and anode material (Pt/BDD), on the degradation of bisphenol A (BPA). By testing different combinations of these factors, the researchers aimed to determine the most significant factors influencing BPA degradation and optimize the conditions for effective EO treatment in cold climate regions.

The 90% removal of the initial BPA concentration took up to 71% (30 min.) longer, and up to 46% more power input was required at 6°C than at 20°C. Temperature and anode material were the two factors that significantly affected the trichloromethane formation. Trichloromethane was formed up to 45% faster at 20°C than at 6°C, and up to 23% faster on Pt than on BDD anodes. The results revealed that, in low average temperature regions, the use of EO for organic pollutant removal is a feasible technology, however, higher energy requirement and consequently higher costs have to be taken into consideration (Ambauen et al. 2020).

Another study (Zaviska et al. 2013) investigated the degradation of chlortetracycline (CTC) hydrochloride in an aqueous medium using Ti = PbO_2 and Ti = IrO_2 anode electrodes through the electrooxidation process. Various operating parameters, including current density, reaction time, and temperature, were analyzed to determine the optimal experimental conditions for CTC degradation. The results showed that the period of treatment time had a more significant impact on CTC removal efficiency, and the effect of current intensity and treatment time were the most influential parameters on the electrochemical oxidation of CTC. The study also found that temperature had a positive effect on CTC removal efficiency, and higher removal efficiencies could be achieved with increasing treatment time at all temperatures studied. The optimal conditions for CTC degradation were found to be Ti = PbO_2, operated at a current intensity of 2.0 A during 49 min. of treatment time and a temperature of 14°C, resulting in 96% of CTC degradation with a total cost of US$3.23 per gram of CTC removed (including only energy and chemical consumption) (Zaviska et al. 2013).

3.2.3 Ultraviolet Technology

Ultraviolet (UV) disinfection is a well-established, cost-competitive technology. UV light can be used to kill some pathogens. More recently, UV disinfection has been attracting a lot of attention due to the discovery of chlorinated disinfection by-products (DBP) and new measurements confirming the effectiveness of UV to inactivate Cryptosporidium. UV is also effective in disinfecting water containing Giardia, *E. coli*, viruses, spores, and other bacteria. Cryptosporidium and Giardia are waterborne pathogenic protozoa implicated in major public health crises throughout the world (Banas et al. 2005; Timmermann et al. 2015).

In the past two years, Sandia researchers have demonstrated up to milliwatt output powers from AlGaN-based light-emitting diodes (LEDs) in the 275–290 nm region of the spectrum. This development has enabled, for the first time, a solid-state light source in the germicidal region of the spectrum (230–300 nm) with output powers that are nearing relevant levels for water purification (Banas et al. 2005; Timmermann et al. 2015).

An assessment of mobile, point-of-use UV purification systems was performed, once again including design considerations for an LED-based system (Banas et al. 2005). The results showed that the optimal operating temperature is 15°C, and the lamp output would fall off rapidly as the lamp temperature migrates from this condition.

Typically these lamps have an efficiency of 25–30%, which is temperature dependent (Banas et al. 2005). For this portable, military-type application, a reasonable goal for the UV LED research program is to develop LEDs with energy conversion efficiencies about 10% at a wavelength as close as possible to 265 nm for maximum germicidal effectiveness (Banas et al. 2005).

UV treatment may include advantages such as: may destroy 99.99% of bacteria, viruses, and protozoa; fast, small, lightweight, economical, and long lasting; non-chemical added; no chemical by-products that cause an irritating smell or toxicity; portable; requires no external electricity source; less time-consuming than chemical treatment (residence time of chemicals between 30 min. and 2 h vs. 90 s of UV irradiation); less time/material consuming than boiling; cool drinking water immediately after application (Banas et al. 2005; Timmermann et al. 2015).

3.2.4 Adsorption Technology

Adsorption technology has emerged as a viable option for water and wastewater treatment, particularly in cold environments where traditional methods may not be effective (Park et al. 2015).

The mobile water treatment package system developed by Park et al. (2015) is a prime example of the potential of adsorption technology in cold environments. The system uses coagulation adsorption for pretreatment and activated carbon (AC) for post-treatment. The results of the study indicate that AC is highly effective in removing a range of pollutants, including COD, TN, NH_3-N, NO_3-N, and DOC (in Table 3.1), thereby improving water quality.

The use of mobile and portable adsorption technology for water/wastewater treatment in cold environments offers several advantages. First, the technology is highly efficient, with the ability to remove a wide range of contaminants from water. Second, it is portable and can be easily transported to remote regions, making it ideal for emergency situations. Additionally, the technology is sustainable, as it requires minimal energy and produces no harmful by-products.

Furthermore, the use of adsorption technology can contribute to the development of a circular economy. Adsorbents, such as activated carbon, can be regenerated and reused, reducing the need for new materials and minimizing waste. This approach not only enhances sustainability but also reduces costs (Park et al. 2015; Lakho et al. 2022). The potential for regeneration and reuse of adsorbents also contributes to the development of a circular economy.

3.3 MOBILE AND PORTABLE TECHNOLOGIES FOR WASTEWATER TREATMENT IN COLD ENVIRONMENTS

3.3.1 Membrane Bioreactor (MBR) Technology

Mobile and portable membrane bioreactor (MBR) technology has been used for water and wastewater treatment in various settings, including remote and cold environments. In these environments, the availability of clean water is often limited, and conventional treatment technologies may not be suitable due to their high energy requirements, large footprint, and limited mobility. MBR technology offers a solution to these challenges by providing a compact, energy-efficient, and versatile option for water and wastewater treatment (Kobayashi et al. 2020; Petropoulos et al. 2021).

MBR technology is a combination of biological treatment and membrane filtration. The biological treatment process uses microorganisms to remove organic and inorganic contaminants from water and wastewater, while the membrane filtration process separates the treated water from the microorganisms and other solids. The membrane filtration process can be performed using either immersed or external membranes. Immersed membranes are submerged in the wastewater, while external membranes are located outside the bioreactor tank (Kobayashi et al. 2020; Petropoulos et al. 2021).

MBR technology offers several advantages over conventional treatment technologies. For example, MBRs can achieve higher removal rates of contaminants, require less space, and produce higher quality effluent. Additionally, MBRs can be operated at a higher biomass concentration, resulting in a smaller bioreactor size and lower energy requirements.

MBR technologies are designed to be compact, lightweight, and transportable. These systems are often used in remote and hard-to-reach locations, such as mining camps, military bases, and disaster relief sites. Mobile and portable MBR systems can be installed on a trailer or a skid, allowing for easy transportation and quick deployment (Kobayashi et al. 2020; Petropoulos et al. 2021).

The feasibility of anaerobic sewage treatment at extremely low temperatures (4°C) was investigated (Petropoulos et al. 2021) using two reactor setups: up-flow anaerobic sludge blanket reactors (UASB) without and with ($AnMBR_{UASB(UF)}$) a membrane. Both reactors were inoculated with seeds derived from sediments that were putatively acclimatized to low temperatures (Petropoulos et al. 2021).

After 40 days of acclimation, both systems met the EU COD effluent standard (<125.0 mg L^{-1}). Initially, the removal efficiency and methane production rate of the AnMBR were slightly higher than those of the UASB. However, over time, both the performance (COD removal and methane production) and the intrinsic capability of the biomass (expressed as cell-specific activity) became similar. The results also suggested that at 4°C, methane can be produced not only through direct acetoclastic methanogenesis but also through acetate oxidation coupled with hydrogenotrophic methanogenesis (Petropoulos et al. 2021)

The use of cold-adapted inocula has been reported to address the issue of prolonged acclimation and unsatisfactory hydrolysis/methanogenesis at low temperatures (Miron et al. 2000; Bowen et al. 2014; Petropoulos et al. 2021).

Safitri (2022) demonstrated the feasibility of an up-flow anaerobic sludge blanket (UASB) system treating municipal wastewater at low temperatures and variable loadings in a long-term application. The system was set at decreasing temperatures (25, 16, 12, 8.5, 5.5, and 2.5°C) and variable organic loading rates from 1.0 $gCOD \cdot l^{-1} \cdot d^{-1}$ up to 15.2 $gCOD \cdot l^{-1} \cdot d^{-1}$, and was investigated over 1025 days. Experiments were performed in two parallel in-house-designed laboratory-scale UASB reactors, which were operated continuously with hydraulic retention time of 16.7 h down to 1.1 h. Stable COD removal efficiencies of 50–70% were achieved at 25°C down to 8.5°C, with loading up to approximately 15.2 $gCOD \cdot l^{-1} \cdot d^{-1}$. COD removal efficiencies were reduced at temperatures below 8.5°C, but significant methane formation was observed even at 2.5°C at reduced loading (up to 5 $gCOD \cdot l^{-1} \cdot d^{-1}$). More than 90% of COD removed was converted to methane, and the methane yield did not change significantly with respect to temperatures. The overall COD balance closed at above 90% of the inlet COD at all operating temperatures and organic loading rates (Safitri 2022).

Temperature has a significant impact on the performance of a reactor, as well as on the structure of the microbial community and the degradation pathway of organic matter. In particular, acetoclastic methanogen plays a critical role in this process, with acetate being the primary precursor for methanogenesis at low temperatures. It is worth noting that regardless of the organic loading rates used, microbial communities can adapt to extremely low temperatures, as low as 2.5°C (Safitri 2022).

However, the UASB reactor is not commonly used in regions with low temperatures. This is mainly due to poor performance, including lower efficiency in removing COD and increased methane production resulting from the low activity of methanogens and hydrolytic bacteria (Petropoulos et al. 2017; Wang 2021). The key limiting factors are increased dissolved methane, decreased activity of methanogens, and hydrolytic bacteria. It proposed that five technologies could reduce the impact of low temperature on the treatment of municipal wastewater by the UASB reactor: denitrifying anaerobic methane oxidation (DAMO), direct interspecies electron transfer (DIET), bioelectrochemical systems (BES), pre-hydrolysis, and degassing membrane (DM). These technologies are promising to be developed into highly efficient engineering strategies for treating municipal wastewater in the UASB reactor at low temperatures in the future (Wang 2021).

He et al. (2018) investigated the nitrogen transformation rates of different nitrogen-loading (20, 30, and 50 mg TN/L) biochar packed reactors (C:N:P = 100:5:1) within 125 days at 5°C. The results showed that high nitrogen loading resulted in an NH_4^+ (TN) removal efficiency decline from 98% (57%) to 83% (29%), with biochar yielding a higher NH_4^+, TN and dissolved organic nitrogen (DON) removal rate than conventional activated sludge. Moreover, all biochar-packed reactors realized a quick start-up by dropping in temperature stage by stage, and the effluent dissolved organic nitrogen (DON) concentrations of R20, R30, and R50 were 0.44±0.18, 0.85±0.35, and 0.66±0.26mg/L, respectively (He et al. 2018).

A study (Elmitwalli, Zeeman, and Lettinga 2001) concerning the feasibility of anaerobic treatment of domestic sewage at low temperature was conducted. The batch tests demonstrated a high biodegradability of domestic sewage at 20°C (74%). Both batch and continuous experiments for the treatment of domestic sewage showed that the removal of suspended solids (SS) prior to anaerobic treatment of domestic sewage not only provides a stable reactor performance but also improves the removal of both colloidal COD (CODcol) and dissolved COD (CODdis). The results of the pretreatment of domestic sewage in an anaerobic filter (AF) and an anaerobic hybrid (AH) reactor showed that the AF reactor is an efficient process for the removal of suspended COD (CODss), viz. 82%, at an hydraulic retention time (HRT) of 4 h and 13°C. The novel AF reactor consists of vertical sheets of reticulated polyurethane foam with knobs, where the biomass was only in attached form. For the treatment of pre-settled sewage at 13°C, the AH reactor, with granular sludge, showed a higher total COD (CODt) removal than the UASB reactor as a result of higher CODcol removal. Therefore, the performance of a two-step system, AF + AH (with granular sludge) reactor, was investigated with different HRTs at 13°C. For optimization of CODss and CODdis, an HRT of 4 + 4 h is needed, whereas for optimization of CODcol removal, an HRT of 4 + 8 h is required. A CODt removal of 71% was achieved with 60% conversion to methane from the removed CODt when the AF + AH system was operated at an HRT of 4 + 8 h at 13°C (Elmitwalli, Zeeman, and Lettinga 2001).

Over the past 30 years, passive treatment of mine-impacted water has become more prevalent. Natural attenuation of contaminants in wetlands, bogs, and swamps has been extensively studied and documented (Sobolewski 1997; MacGregor 2002). Passive Treatment Systems (PTSs) consist of constructed wetlands (CWs), bioreactors, or some combination and/or variation on both of these components. Other types of PTSs exist, such as in-pit or in-pond treatment systems (Ness, Janin, and Stewart 2014). Natural wetlands have been treating mine-impacted water in cold climates since mining activity began (Ness, Janin, and Stewart 2014).

Some of the challenges facing PTSs in cold climates include hydraulic failure due to freezing in winter, reduced microbial activity, and increased water flows during spring melt. Many adaptations have been developed to overcome these challenges, such as insulating and/or burying PTSs, maintaining constant flows through the use of bypasses and overflows, and providing carbon sources to feed bacteria during times of slower organic matter decomposition, where relevant cold climate adaptations have been used (Ness, Janin, and Stewart 2014).

They have been used extensively in cold climates for municipal wastewater treatment (Mander and Jenssen 2003) and in several northern settings for mine drainage (Sobolewski 1996; Nyquist and Greger 2009; Ness, Janin, and Stewart 2014).

Bioreactors have been used at numerous mine sites in cold climates with limited access and power due to low construction and operating costs (Schmidtova and Baldwin 2011) and their ability to work year-round in sub-zero temperatures (Ness, Janin, and Stewart 2014).

PTSs in northern environments are often located in remote locations with limited access and power, short growing seasons, and cold winter temperatures, conditions that can present challenges for PTS design. The main problems faced by PTSs in cold climates are hydraulic failure due to freezing, ice formation, and insufficient biological activity at low temperatures (Mander and Jenssen 2003; Ness, Janin, and Stewart 2014).

3.3.2 Electrochemical Treatment

A research project (Vialkova and Glushchenko 2021) was carried out to enhance the sewage treatment systems in small, isolated settlements in the Arctic region, aiming to prevent industrial pollution of the surrounding environment. In the course of the research, the stages of coagulation, filtration, oxidation, and chemical precipitation of wastewater pollution were tested, respectively. The best results were obtained at the third stage of the chemical precipitation method of treatment, when it was possible to reduce the concentration of NH_4^+ by 99.5%, SS by 99.9%, and COD by 96.6%. The temperature of the wastewater samples did not affect the processes significantly (Vialkova and Glushchenko 2021).

3.4 CASE STUDIES

Numerous PTSs have been constructed in cold climates at various scales. For example, at the Smoky River Coal Mine in British Columbia, Canada, two 5-m^3 pilot-scale bioreactors were constructed in parallel to treat Se-contaminated water in an area where access, space, and power were severely restricted (Sobolewski, 2010). The bioreactors were fed with ethylene glycol as a liquid carbon source. The carbon source was used to sustain microbial processes (Sobolewski, 2010). The Se removed efficiency was 97% from September to November 2008, even when temperatures were 2°C. The results showed that the addition of a carbon source may allow anaerobic bioreactors to keep a high efficiency on northern mine sites under winter climate conditions (Ness, Janin, and Stewart 2014).

At the Kongens & Folldal Mines in Norway, bioreactors have been used for the treatment of acidic mine drainage at both active and abandoned mines in cold climates (Ettner 2007). A first pilot bioreactor system was constructed at the Kongens Mine in northern Norway in 1999. It consisted of a pretreatment limestone drain and four anaerobic cells filled with organic waste materials, rotted hay, sawdust, tree clippings, and $CaCO_3$.

A new bioreactor system was constructed in 2002 using rotted hay, sand, and coarse birch tree materials. In this reactor, up to 85% Cu, 48% Zn, 98% Cd, 82% Pb, 71% Cr, and 24% Ni were removed in this second pilot process, however Cu removal was significantly reduced in winter months. This reduction was correlated with a drastic increase in the flow rate in the bioreactor (Ettner 2007).

To overcome this issue and avoid high flow variations, a full-scale bioreactor was constructed in 2006 with four 8 × 1.5-m basins filled with rotten hay, tree clippings, shell sand, and gypsum to act as an extra sulfate source (Ettner 2007). The study indicated that cold winter temperatures did not prevent inoculation of the bioreactors by microorganisms, and development of anaerobic conditions, low oxidation/reduction potential (Eh), and signs of sulfide precipitates were reported (Ness, Janin, and Stewart 2014).

3.5 CHALLENGES IN THE TECHNOLOGIES

Research has shown that ponds and wetlands can effectively treat wastewater in arctic regions (Ness et al. 2014). However, design guidelines for these passive treatment systems need to be refined to account for the unique climate conditions and the short summer treatment season (Daley et al. 2015). While passive treatment systems appear to be best suited for the North, there is still potential for improving the operation of the small number of mechanical treatment systems currently in use.

Wastewater treatment in northern regions is facing an emerging issue of dealing with pharmaceuticals and other refractory contaminants (per- and polyfluoroalkyl substances (PFAS), micro-plastics, nano-plastics, etc.) being released into the environment. Pharmaceuticals have low biodegradability and persist for long periods in cold waters, with antibiotics being a particular concern due to their potential to disrupt aquatic organisms and accumulate in the food chain. Addressing this issue may require solutions beyond the capabilities of existing treatment systems. As most pharmaceutical residues in wastewater come from human excreta, using source-separating technologies such as dry toilets to separate toilet waste from other wastewater (e.g. laundry or bathing water) could be a cost-effective option for arctic communities (Daley et al. 2015).

As industrial activity increases in cold climate regions, there is a growing need for the research, development, and implementation of cost-effective systems that can ensure long-term environmental protection.

Low-temperature anaerobic treatment of domestic wastewater, even with the use of psychrophiles, is limited by poor lipid degradation. Therefore, further research is required to identify the true lipid degraders in these systems.

Effective wastewater management in arctic regions requires approaches that prioritize both environmental and health protection while being appropriate for the unique cold climate conditions (Daley 2016). Currently, the most successful approach involves using simple yet robust systems. However, given the rapidly changing conditions in the arctic, it is crucial for communities and other stakeholders to remain informed about emerging challenges and innovative solutions.

Mobile and portable membrane reactor (MBR) technologies also have some limitations. These systems require a constant supply of electricity to operate, which can be challenging in remote locations with limited access to power. Additionally, MBRs may require frequent maintenance and monitoring to ensure proper operation and prevent equipment failure. However, with the development of advanced control systems and renewable energy sources, these limitations can be minimized, making MBRs a viable option for water and wastewater treatment in cold environments (Kobayashi et al. 2020; Petropoulos et al. 2021).

3.6 CONCLUSIONS AND PERSPECTIVES

In today's world, the need for immediate and efficient solutions for water and wastewater treatment has become increasingly critical, particularly during emergencies, unplanned plant shutdowns, or in remote areas where access to traditional water infrastructure is limited. Fortunately, mobile and portable technologies designed specifically for water and wastewater treatment in cold environments have emerged as a transformative solution, providing rapid and flexible responses to address these pressing challenges.

The significance of these mobile and portable technologies cannot be overstated, especially in situations where time is of the essence. During emergencies, such as natural disasters or industrial accidents, access to safe and clean water is paramount for the survival and well-being of affected communities. Traditional water treatment facilities may be compromised or inaccessible in these scenarios, making mobile and portable systems essential lifelines.

Moreover, in remote and isolated regions, these technologies offer a lifeline to communities that lack access to centralized water treatment infrastructure. They can be rapidly deployed to provide immediate relief and support until more permanent solutions can be established.

One of the key advantages of these systems is their versatility and adaptability. By simplifying process configurations and utilizing modularization, these technologies can be tailored to suit the unique characteristics of various water sources. This adaptability ensures that unit processes can be selectively combined based on raw water properties, allowing for automatic selection and operation. Furthermore, the prospect of self-generated power consumption in the future makes these systems more sustainable and less reliant on external energy sources.

The mobile water treatment package system exemplifies this versatility, as it can be used for both small-scale water supplies and emergency situations. Its mobility allows it to be quickly transported to where it is needed the most, providing a stable and reliable emergency water supply during critical times. The ease of deployment and operation makes it an invaluable asset in any water-related crisis.

Looking ahead, there are exciting prospects for the continued improvement and expansion of mobile and portable water treatment technologies. Modularization is set to play a crucial role in this evolution, allowing for even greater customization and efficiency. By applying only the necessary treatment processes based on specific water source characteristics, resource utilization can be optimized, resulting in more cost-effective and environmentally friendly solutions.

Investing in research and development in this field will be pivotal in advancing these technologies further. Collaborations between experts in water treatment, environmental engineering, and material sciences can lead to groundbreaking innovations that will revolutionize how we respond to water-related challenges in cold environments.

By embracing modularization and fostering continuous research, these technologies are ensured to continue to improve, paving the way for a more resilient and sustainable approach to water management globally. Ultimately, these innovative solutions will play a pivotal role in securing access to safe and clean water for generations to come.

REFERENCES

Ambauen, Noëmi, Jens Muff, Franz Tscheikner-Gratl, Thuat T Trinh, Cynthia Halle, and Thomas Meyn. 2020. "Application of electrochemical oxidation in cold climate regions–Effect of temperature, pH and anode material on the degradation of Bisphenol A and the formation of disinfection by-products." *Journal of Environmental Chemical Engineering* 8 (5):104183.

Banas, Michael Anthony, Mary Hagerott Crawford, Douglas Scott Ruby, Michael P Ross, Jeffrey Scott Nelson, Andrew Alan Allerman, and Ray Boucher. 2005. Final LDRD report: ultraviolet water purification systems for rural environments and mobile applications. Sandia National Laboratories (SNL), Albuquerque, NM, and Livermore, CA.

Bowen, Emma J, Jan Dolfing, Russell J Davenport, Fiona L Read, and Thomas P Curtis. 2014. "Low-temperature limitation of bioreactor sludge in anaerobic treatment of domestic wastewater." *Water Science and Technology* 69 (5):1004–1013.

Daley, Kiley. 2016. "Wastewater Management: Northern Solutions for Northern Conditions." PhD Candidate, Dalhousie University, https://www.canada.ca/en/polar-knowledge/publications/polarleads/vol1-no2-2016.html

Daley, Kiley, Heather Castleden, Rob Jamieson, Chris Furgal, and Lorna Ell. 2015. "Water systems, sanitation, and public health risks in remote communities: Inuit resident perspectives from the Canadian arctic." *Social Science & Medicine* 135:124–132. doi: https://doi.org/10.1016/j.socscimed.2015.04.017.

Elmitwalli, T, Gr Zeeman, and G Lettinga. 2001. "Anaerobic treatment of domestic sewage at low temperature." *Water Science and Technology* 44 (4):33–40.

Ettner, David C. 2007. "Passive mine water treatment in Norway." MWA. Proceedings of the International Mine Water Association 2007 Symposium. Cagliari (Italy).

Faust, Samuel D, and Osman M Aly. 2018. *Chemistry of water treatment*: CRC press.

Gunnarsdóttir, Ragnhildur, Petter Deinboll Jenssen, Pernille Erland Jensen, Arne Villumsen, and Roland Kallenborn. 2013. "A review of wastewater handling in the Arctic with special reference to pharmaceuticals and personal care products (PPCPs) and microbial pollution." *Ecological Engineering* 50:76–85.

He, Su, Lili Ding, Yao Pan, Haidong Hu, Lin Ye, and Hongqiang Ren. 2018. "Nitrogen loading effects on nitrification and denitrification with functional gene quantity/transcription analysis in biochar packed reactors at 5 C." *Scientific Reports* 8 (1):1–9.

Kabeel, A. E., and Emad M. S. El-Said. 2015. "Water production for irrigation and drinking needs in remote arid communities using closed-system greenhouse: A review." *Engineering Science and Technology, an International Journal* 18 (2):294–301. doi: https://doi.org/10.1016/j.jestch.2014.12.003.

Kobayashi, Yumi, Nicholas J. Ashbolt, Evan G. R. Davies, and Yang Liu. 2020. "Life cycle assessment of decentralized greywater treatment systems with reuse at different scales in cold regions." *Environment International* 134:105215. doi: https://doi.org/10.1016/j.envint.2019.105215.

Lakho, Fida Hussain, Hong Quan Le, Flor Mattheeuws, Wouter Igodt, Veerle Depuydt, Joachim Desloover, Diederik P. L. Rousseau, and Stijn W. H. Van Hulle. 2021. "Decentralized grey and black water reuse by combining a vertical flow constructed wetland and membrane based potable water system: Full scale demonstration." *Journal of Environmental Chemical Engineering* 9 (1):104688. doi: https://doi.org/10.1016/j.jece.2020.104688.

Lakho, Fida Hussain, Hong Quan Le, Fréderic Van Kerkhove, Wouter Igodt, Veerle Depuydt, Joachim Desloover, Diederik P. L. Rousseau, and Stijn W. H. Van Hulle. 2020. "Water treatment and re-use at temporary events using a mobile constructed wetland and drinking water production system." *Science of The Total Environment* 737:139630. doi: https://doi.org/10.1016/j.scitotenv.2020.139630.

Lakho, Fida Hussain, Asif Qureshi, Wouter Igodt, Hong Quan Le, Veerle Depuydt, Diederik P. L. Rousseau, and Stijn W. H. Van Hulle. 2022. "Life cycle assessment of two decentralized water treatment systems combining a constructed wetland and a membrane based drinking water production system." *Resources, Conservation and Recycling* 178:106104. doi: https://doi.org/10.1016/j.resconrec.2021.106104.

Letterman, Raymond D. 1999. *Water quality and treatment: A handbook of community water supplies*: McGraw-hill New York.

MacGregor, Dylan B. 2002. *Natural attenuation of acqueous zinc in shallow soils over permafrost downslope of Galkeno 300 Mine Adit, United Keno Hill Mines, Central Yukon*: University of British Columbia.

Mander, Ülo, and PD Jenssen. 2003. *Constructed wetlands for wastewater treatment in cold climates*: WIT Press Southampton.

Miron, Yehuda, Grietje Zeeman, Jules B Van Lier, and Gatze Lettinga. 2000. "The role of sludge retention time in the hydrolysis and acidification of lipids, carbohydrates and proteins during digestion of primary sludge in CSTR systems." *Water Research* 34 (5):1705–1713. doi:

Ness, I, A Janin, and K Stewart. 2014. "Passive treatment of mine impacted water in cold climates: A review." *Yukon Research Centre, Yukon College*:56.

Northcott, Kathy A, Ian Snape, Michael A Connor, and Geoffrey W Stevens. 2003. "Water treatment design for site remediation at Casey Station, Antarctica: Site characterisation and particle separation." *Cold Regions Science and Technology* 37 (2):169–185.

Northcott, Kathy A, Ian Snape, Peter J Scales, and Geoff W Stevens. 2005. "Contaminated water treatment in cold regions: An example of coagulation and dewatering modelling in Antarctica." *Cold Regions Science and Technology* 41 (1):61–72.

Northcott, Kathy A., Penny Woodberry, Ian Snape, and Geoff W. Stevens. 2007. "Water treatment to prevent contaminant dispersal during remediation of cold regions contaminated sites." *Cold Regions Science and Technology* 48 (2):92–104. doi: https://doi.org/10.1016/j.coldregions.2006.07.003.

Nyquist, Johanna, and Maria Greger. 2009. "A field study of constructed wetlands for preventing and treating acid mine drainage." *Ecological Engineering* 35 (5):630–642.

Park, Young Kyu, Ju-Suk An, Jiyoung Park, and Hyun Je Oh. 2015. "Development of mobile water treatment package system for emergency water supply." *International Journal of Structural and Civil Engineering Research* 4 (3):296–300.

Petropoulos, Evangelos, Jan Dolfing, Russell J. Davenport, Emma J. Bowen, and Thomas P. Curtis. 2017. "Developing cold-adapted biomass for the anaerobic treatment of domestic wastewater at low temperatures (4, 8 and 15°C) with inocula from cold environments." *Water Research* 112:100–109. doi: https://doi.org/10.1016/j.watres.2016.12.009.

Petropoulos, Evangelos, Burhan Shamurad, Shamas Tabraiz, Yongjie Yu, Russell Davenport, Thomas P Curtis, and Jan Dolfing. 2021. "Sewage treatment at 4° C in anaerobic upflow reactors with and without a membrane–performance, function and microbial diversity." *Environmental Science: Water Research & Technology* 7 (1):156–171.

Safitri, Anissa Sukma. 2022. "Low-temperature anaerobic wastewater treatment by granulated biomass." Stavanger : University of Stavanger, 2022 (PhD thesis UiS, no. 648)

Schmidtova, Jana, and Susan A Baldwin. 2011. "Correlation of bacterial communities supported by different organic materials with sulfate reduction in metal-rich landfill leachate." *Water Research* 45 (3): 1115–1128.

Smith, Daniel W. 1996. "Cold regions utilities monograph." American Society of Civil Engineers; 3rd edition, p. 780

Sobolewski, André. 1996. "Metal species indicate the potential of constructed wetlands for long-term treatment of metal mine drainage." *Ecological Engineering* 6 (4):259–271.

Sobolewski, André. 1997. "The capacity of natural wetlands to ameliorate water quality: a review of case studies." Conference: 4. International Conference on Acid Rock Drainage, Vancouver (Canada), 31 May–6 Jun 1997; Other Information: DN: MEND Secretariat, CANMET, Natural Resources Canada, 555 Booth Street, Ottawa, ON, K1A 0G1, Canada; PBD: 1997; Related Information: Is Part Of Fourth international conference on acid rock drainage, proceedings. Volume IV; PB: 4 vols., 1937 p.; https://www.osti.gov/etdeweb/biblio/321799

Sobolewski, A. 2010. Benefits of using liquid carbon sources for passive treatment systems, in C. Wolkersdorfer (ed), Proceedings of International Mine Water Association Conference: Mine Water and Innovative Thinking. Sydney (NS): 279–282.

Timmermann, Lisa F, Klaus Ritter, David Hillebrandt, and Thomas Küpper. 2015. "Drinking water treatment with ultraviolet light for travelers–Evaluation of a mobile lightweight system." *Travel Medicine and Infectious Disease* 13 (6):466–474.

Vialkova, Elena, and Ekaterina Glushchenko. 2021. "Wastewater treatment in remote arctic settlements." *Water* 13 (7):919.

Wang, Zhenyu. 2021. "Effect of low temperature on the performance of UASB reactor for municipal wastewater treatment: Recent advances and future perspectives." *IOP Conference Series: Earth and Environmental Science* 621 (1):012103. doi: 10.1088/1755-1315/621/1/012103.

Zaviska, François, Patrick Drogui, Jean-François Blais, Guy Mercier, and Samuel De La Rochebrochard d'Auzay. 2013. "Electrochemical oxidation of chlortetracycline using Ti/IrO 2 and Ti/PbO 2 anode electrodes: Application of experimental design methodology." *Journal of Environmental Engineering* 139 (6):810–821.

Zhang, Wenchang, Yu Xia, Zhaotong Wen, Wenxia Han, Shaofu Wang, Yiping Cao, Rong-Xiang He, Yumin Liu, and Bolei Chen. 2021. "Enhanced adsorption-based atmospheric water harvesting using a photothermal cotton rod for freshwater production in cold climates." *RSC Advances* 11 (56): 35695–35702.

4 Potential Decentralized Anaerobic Treatment of Wastewater in Cold Environments

Song Yan[a], Patrick Drogui[a], Rajeshwar D. Tyagi[b,c], and Jonathan W.C. Wong[b]
[a] Université du Québec, Québec, Canada
[b] Dongguan University of Technology, Dongguan, China
[c] BOSK Bioproducts, Québec, Canada

4.1 INTRODUCTION

Anaerobic wastewater treatment has been successfully demonstrated for more than a century. Various processes have been developed, including anaerobic suspended growth, up-flow and down-flow anaerobic attached growth, fluidized-bed attached growth, anaerobic lagoons, membrane separation anaerobic processes, and the up-flow anaerobic sludge blanket (UASB) process. The UASB process is a very typical process. Moreover, several modern high-rate anaerobic wastewater treatment systems have been developed for various types of wastewaters, such as the expanded granular sludge bed (EGSB) reactor, anaerobic baffled reactor (ABR), and anaerobic membrane bioreactors (Tchobanoglus, Burton, and Stensel 2003; Van Lier, Mahmoud, and Zeeman 2008).

Anaerobic treatment presents a multitude of significant advantages, including low sludge production, low energy requirements, and the potential for energy recovery. Anaerobic treatment can be conducted in technically plain systems, and the process can be applied at any scale and at almost any place; this makes it a potential decentralized anaerobic treatment of wastewater in cold environments. With its wide range of applications, anaerobic treatment is considered a well-established method in environmentally sustainable development and a promising technology for the future (Tchobanoglus, Burton, and Stensel 2003; Van Lier, Mahmoud, and Zeeman 2008).

It has been observed by many researchers that temperature has a strong effect on the maximum substrate utilization rates of microorganisms in the process (Lew et al. 2004; O'Reilly et al. 2009; McKeown et al. 2012; Bowen et al. 2014; Madden et al. 2014; Petropoulos et al. 2017; Deng et al. 2020; Bashiri et al. 2022). The treatment of decentralized wastewater in rural areas, particularly in cold climate conditions, remains a challenge.

In this chapter, recent advances and developments in potential decentralized anaerobic treatment of wastewater in cold environments are presented. The fundamentals of anaerobic treatment, the role of extremophilic anaerobes, and the development of anaerobic processes are highlighted. Moreover, anaerobic biochemistry, microbiology, and industrial and technological aspects are also emphasized. Cost estimation and techno-economic feasibility are illustrated. In addition, the challenges of the processes in cold conditions are described.

DOI: 10.1201/9781003376415-5

4.2 FUNDAMENTALS OF ANAEROBIC TREATMENT

The generally accepted definition of anaerobic treatment is a microbiologically mediated process during which organic carbon present in biopolymers and other degradable compounds is converted to its most reduced form (methane) and its most oxidized form (carbon dioxide) in the absence of oxygen (anaerobic conditions).

This energy-efficient process reliably removes biochemical oxygen demand (BOD), chemical oxygen demand (COD), and total suspended solids (TSS) from wastewater, thus resulting in an effluent with lower BOD, COD, and/or TSS, as well as biogas byproducts (mainly methane and carbon dioxide).

Anaerobic treatment is typically applied for streams with high concentrations of organic material (measured as high BOD, COD, or TSS), often prior to aerobic treatment.

The complex, biologically mediated anaerobic digestion events are far from being understood in detail (Madsen, Holm-Nielsen, and Esbensen 2011).

The microbiology of anaerobic digestion is complicated, since it involves several bacterial groups, each performing a separate task in the overall degradation process. So far, up to nine steps have been identified during the anaerobic conversion of organic matter. For simplicity, the anaerobic process can be characterized by four sequential processes: hydrolysis, acidogenesis, acetogenesis, and methanogenesis. Three major bacterial groups are involved: the hydrolytic-fermentative bacteria that hydrolyze and convert the organic compounds to volatile fatty acids with the simultaneous production of hydrogen (H_2) and carbon dioxide (CO_2), the acetogenic bacteria that convert the aforementioned acids to acetic acid; and, finally, the methanogenic bacteria that produce methane, either from acetate or from H_2 and CO_2 (Lew et al. 2004; Chu, Yang, and Zhang 2005; Van Lier, Mahmoud, and Zeeman 2008; Lew et al. 2011; Mukhtar and Aslam 2020). These byproducts can be reclaimed for use as fuel, while the wastewater can be routed for further treatment and/or discharge (Gavala, Angelidaki, and Ahring 2003).

During the first hydrolysis process, hydrolytic bacterial species degrade a particular type of macromolecule. The hydrolysis rate depends on the temperature, pH, hydrolyzing mass concentration, and size/type of particulate organic matter (Visvanathan and Abeynayaka 2012).

In the second step of acidogenesis, soluble sugars, amino acids, and fatty acids are converted into short chain volatile fatty acids (VFAs: formic, butyric, propionic, and acetic acid), alcohols, aldehydes, hydrogen, and carbon dioxide. Acidogenic bacteria are rapid growing micro-species and are affected by environmental conditions (Visvanathan and Abeynayaka 2012).

Acetogenesis is the third step. Acetogenic bacteria convert VFAs, alcohols, and hydrogen into acetate (Xie et al. 2014). The process of acetogenesis is led by two groups of bacteria: oxidation of butyrate, propionate, and long chain fatty acids to acetate (obligate H_2 producing bacteria), and intake of H_2 and CO_2 to produce acetate by homoacetogens (Ketheesan and Stuckey 2015, Xie et al. 2014, Hussain and Dubey 2017).

The last step is methanogenesis, in which methane is produced. Methane production is a two-way process in anaerobic digestion. In the first way, methane is produced from acetate by acetoclastic methanogens such as Methanosarcina and Methanosaeta (Ketheesan and Stuckey 2015). Approximately 70% of methane is produced via acetoclastic methanogenesis. The second way of methane production is through hydrogenotrophic methanogens, which converts carbon dioxide and hydrogen to methane and carbon dioxide (Maaz et al. 2019)

4.3 ANAEROBIC TREATMENT IN COLD CLIMATE CONDITIONS

Anaerobic digestion is strongly influenced by temperature and can be grouped under one of the following categories (Hulshoff-Pol 1998): psychrophilic (0–20°C), mesophilic (20–42°C), and thermophilic (42–75°C). The details of the bacterial processes in all three temperature ranges are well established, though a large section of the reported work deals with the mesophilic operation.

Changes in temperature are well resisted by anaerobic bacteria, as long as they do not exceed the upper limit as defined by the temperature at which the decay rate begins to exceed the growth rate. In the mesophilic range, the bacterial activity and growth decreases by one-half for each 10°C drop below 35°C (Hulshoff-Pol 1998). Thus, for a given degree of digestion to be attained, the lower the temperature, the longer the digestion time. The effect of temperature on the first stage of the digestion process (hydrolysis and acidogenesis) is not very significant. The second and third stages of decomposition can only be performed by certain specialized microorganisms (acetogenic and methanogenic bacteria) and, thus, these are much more sensitive toward temperature change. However, an important characteristic of anaerobic bacteria is that their decay rate is very low at temperatures below 15°C. Thus, it is possible to preserve the anaerobic sludge for long periods without losing much of its activity. This is especially useful in the anaerobic treatment of wastewater from seasonal industries such as sugar mills (Saleh and Mahmood 2004).

Various investigations have been carried out to overcome the obstacles of anaerobic treatment of domestic sewage at lower temperatures (Table 4.1). Besides operational aspects such as sludge characteristics and up-flow velocities, the technical design of the overall process influences the performance (Halalsheh 2002, Seghezzo 2004).

One of the current solutions for treatment of raw domestic sewage at low temperature is a two-step UASB system, where in the first step, mainly the removal of suspended solids and partial hydrolysis, acidification occurs, while in the second step the hydrolyzed organic matter is converted to biogas (Halalsheh 2002, Seghezzo 2004). However, results showed similar performance of the two-step reactor in comparison to a one-step system, due to lower removal efficiencies and lower sludge retention time (SRT) in the second stage. Typical problems of highly loaded UASB reactors like sludge flotation and washout of active biomass were observed, mainly at temperatures below 20°C. Moreover, large amounts of sludge accumulated in the primary reactor and needed to be further treated in a separate system.

Mahmoud et al. (2004) investigated a combined UASB-CSTR (completely stirred tank reactor) digester system, where the accumulated sludge from the UASB reactor was then directed to a CSTR-digester, operating at 35°C, for further treatment. Results showed that the UASB-CSTR digester had a better performance than a single-stage UASB reactor at 15°C, with a COD removal of 66% versus 44% (Lew et al. 2011).

The use of an anaerobic process for domestic wastewater treatment would achieve a lower carbon footprint as it eliminates aeration and generates methane. Among several anaerobic treatment processes, high-rate anaerobic digesters receive great attention due to their high loading capacity and chemical oxygen demand removal rate.

The up-flow anaerobic sludge blanket reactor (UASB) is gaining wide acceptance among several anaerobic processes. However, its application is still limited to industrial wastewater treatment. There are some unresolved problems to be accepted in developing countries with lower temperature conditions for domestic wastewater treatment. Several studies have been carried out for the performance of UASB reactors, but there is still a lack of updated information, especially on the issue of low-temperature domestic wastewater treatment. Considering the gravity of the issue, an attempt has been made to compile updated information so as to help engineers, researchers, and practitioners in the selection of reactors and future prospects of research (Lohani, Chhetri, and Khanal 2015).

Direct anaerobic treatment in the main line of a wastewater treatment plant is a promising way of recovering chemical energy, even in moderate climates. Full-scale applications of up-flow UASB reactors are common in warm climates and have recently begun to appear in areas with colder climates. Several new types of reactors optimized for low-temperature treatment have been proposed, such as UASB-Digester and anaerobic membrane bioreactor (AnMBR). However, long-term operation of such systems treating real wastewater at winter temperatures is needed to assess their suitability.

The operation of a laboratory-scale UASB reactor (1.9 L) was evaluated before and after the installation of ultrafiltration membrane for solids separation from the effluent. The laboratory model

TABLE 4.1
Summary of Performance of Various Anaerobic Wastewater Treatment in Cold Conditions

Reactor	Temperature (°C)	Organic Loading Rate (g COD/[L·d])	Hydraulic Retention Time	COD Removal (%)	Biogas Production Methane Content (%)	Reference
Laboratory-scale UASB reactor (1.9 L)	15	1.4 COD/[L·d]	9 h	64%	63%±17%	(Hejnic et al. 2016)
AnMBR	15	1.6 COD/[L·d]	10 h	85%	64%±2%	(Hejnic et al. 2016)
Two-anaerobic hybrid (AH) septic tanks (each 0.575 m^3)			2.5 days	Removal efficiencies in the two-step AH-septic tank at 5 days HRT and 13°C were 94, 98, 74, and 78% for total COD, suspended COD, colloidal COD, and dissolved COD		(Elmitwalli et al. 2003)
One-step AH septic tank	13°C		5.5–7.5 days	87%		(Elmitwalli et al. 2003)
Expanded granular sludge bed (EGSB)	4–5°C 10–12°C	10–12 kg COD per cubic meter reactor per day		90%		(Lettinga, Rebac, and Zeeman 2001)
A two-step (UASB) + EGSB	13°C			50%		(Lettinga, Rebac, and Zeeman 2001)
Anaerobic filter (AF) combined with an anaerobic hybrid reactor (AH)	13°C			70%		(Lettinga, Rebac, and Zeeman 2001)
Bench-scale anaerobic membrane bioreactor (AnMBR) equipped with submerged flat-sheet microfiltration membranes	15°C			92 ± 5%		(Smith, Skerlos, and Raskin 2013)
Actual DWW in bench-scale anaerobic membrane bioreactor (AnMBR) equipped with submerged flat-sheet microfiltration membranes	15°C			69 ± 10%		(Smith, Skerlos, and Raskin 2013)

(Continued)

TABLE 4.1 (Continued)
Summary of Performance of Various Anaerobic Wastewater Treatment in Cold Conditions

Reactor	Temperature (°C)	Organic Loading Rate (g COD/[L·d])	Hydraulic Retention Time	COD Removal (%)	Biogas Production Methane Content (%)	Reference
An up-flow anaerobic sludge blanket (UASB) reactor and a hybrid UASB-filter	28, 20, 14, and 10°C	3.5–5.7 h		Reactor types at 28 and 20°C, 82 and 72% respectively. At 14 and 10°C the UASB reactor showed a better COD removal (70% and 48%, respectively) than the hybrid reactor (60% and 38%).		(Lew et al. 2004)
Expanded granular sludge bed (EGSB) reactor coupled with hollow fiber membrane filtration	11–25°C	3.5–5.7 h		85–96% of (COD) and 83–94% of total organic carbon (TOC). At 11°C, increasing the total COD removal was increased from 76 to 81%.		(Chu, Yang, and Zhang 2005)
Each (3.84 liter) one UASB and two anaerobic hybrid (AH) reactors with small sludge granules	13°C	8 h	1.8 m/d	The use of sheets in the AH reactors: 87% In shallow AH reactors containing small sludge granules removed: 64% UASB: 60%		(Elmitwalli et al. 1999)

treated wastewater at 15°C for more than 850 days. The organic loading rate of the UASB reactor and AnMBR was 1.4 and 1.6 g COD/[L·d], respectively, while the hydraulic retention time was around 9 and 10 h. Energy recovery from the wastewater in gaseous methane was 4% and 6% in the UASB reactor and AnMBR, respectively. The methane content was 63% ± 17% and 64% ± 2%, respectively. In the UASB reactor, 64% of the influent COD was removed in the long term. The AnMBR ran at average 85% COD removal. Results showed that AnMBR could deliver more stable effluent quality for wastewater treatment even at low (15°C) temperatures. However, even the AnMBR effluent required further post-treatment. Thus, simpler and less energy demanding UASB appeared more relevant for wastewater pretreatment under European conditions (Hejnic et al. 2016).

The decentralized treatment of concentrated sewage (about 3600 mgCOD/L) at low temperature was investigated in a two-step anaerobic system: two anaerobic hybrid (AH) septic tanks (each 0.575 m^3). The two reactors were placed in a temperature-controlled room and the hydrolic retention time (HRT) was 2.5 days for each reactor. The system was fed with concentrated domestic sewage, mainly blackwater from about 40 toilets flushed with only 4 L of water and a limited amount of greywater. The system showed high removal efficiency for the different COD fractions. Mean removal efficiencies in the two-step AH septic tank at five days HRT and 13°C were 94, 98, 74, and 78% for total COD, suspended COD, colloidal COD, and dissolved COD, respectively. The results of short-run experiments indicated that the presence of reticulated polyurethane foam (RPF) media in the AH septic tank improved the removal of suspended COD by 22%. The first AH septic tank was full of sludge after four months of operation due to the high removal of particulate COD and the limited hydrolysis at low temperature conditions. Therefore, a simple mathematical model was developed based on Anaerobic Digestion Model No.1 (ADM1) (the International Water Association (IWA) model in 2002). Based on the experimental results and the mathematical model, only a one-step AH septic tank is required. An HRT of 5.5–7.5 days is needed for that one-step AH septic tank to treat concentrated sewage at a low temperature of 13°C. Such a system can provide a total COD removal as high as 87% and will be full of sludge after a period of more than a year (Elmitwalli et al. 2003).

Psychrophilic anaerobic treatment is an attractive option for wastewaters that are discharged at moderate to low temperatures. The expanded granular sludge bed (EGSB) reactor has been shown to be a feasible system for anaerobic treatment of mainly soluble and pre-acidified wastewater at temperatures of 5–10°C. An organic loading rate (OLR) of 10–12 kg COD per cubic meter reactor per day can be achieved at 10–12°C with a removal efficiency of 90%. Further improvement might be obtained by a two-module system in series. Stabile methanogenesis was observed at temperatures as low as 4–5°C. The specific activity of the mesophilic granular sludge was improved under psychrophilic conditions, which indicated that there was growth and enrichment of methanogens and acetogens in the anaerobic system. Anaerobic sewage treatment is a real challenge in moderate climates because sewage belongs to the "complex" wastewater category and contains a high fraction of particulate COD. A two-step system consisting of either a UASB reactor combined with an EGSB reactor or an anaerobic filter (AF) combined with an anaerobic hybrid reactor (AH) was successful for anaerobic treatment of sewage at 13°C with a total COD removal efficiency of 50% and 70%, respectively (Lettinga, Rebac, and Zeeman 2001).

Particularly under temperate climate conditions, many low- and medium-strength wastewaters are discharged at low ambient temperatures, including domestic wastewater and a large variety of industrial wastewaters, such as those from bottling, malting, brewery, and soft drink manufacturing plants.

An AnMBR equipped with submerged flat-sheet microfiltration membranes was operated at psychrophilic temperature (15°C) treating simulated and actual domestic wastewater (DWW). COD removal during simulated DWW operation averaged 92 ± 5%, corresponding to an average permeate COD of 36 ± 21 mg/L. Dissolved methane in the permeate stream represented a substantial fraction (40–50%) of the total methane generated by the system due to methane solubility at psychrophilic temperatures and oversaturation relative to Henry's law. During actual DWW operation, COD removal averaged 69 ± 10%. The permeate COD and five-day biochemical oxygen demand

(BOD_5) averaged 76 ± 10 mg/L and 24 ± 3 mg/L, respectively, indicating compliance with the U.S. Environmental Protection Agency's standard for secondary effluent (30 mg/L BOD_5). Membrane fouling was managed using biogas sparging and permeate backflushing, and a flux greater than 7 Liter/m^2/h (LMH) was maintained for 30 days. Comparative fouling experiments suggested that the combination of the two fouling control measures was more effective than either fouling prevention method alone. A UniFrac-based comparison of bacterial and archaeal microbial communities in the AnMBR and three different inocula using pyrosequencing targeting 16S rRNA genes suggested that mesophilic inocula are suitable for seeding psychrophilic AnMBRs treating low-strength wastewater. Overall, the research described relatively stable COD removal, acceptable flux, and the ability to seed a psychrophilic AnMBR with mesophilic inocula, indicating future potential for the technology in practice, particularly in cold and temperate climates where DWW temperatures are low during part of the year (Smith, Skerlos, and Raskin 2013).

The performance of a UASB reactor and a hybrid UASB-filter reactor was investigated (Lew et al. 2004) and compared for the treatment of domestic wastewater at different operational temperatures (28, 20, 14, and 10°C) and loading rates. For each temperature studied, a constant CODt removal was observed as long as the up-flow velocity was lower than 0.35 m/h. At these up-flow velocities, similar removals were observed for both reactor types at 28°C and 20°C, 82% and 72%, respectively. However, at 14°C and 10°C the UASB reactor showed a better COD removal (70% and 48%, respectively) than the hybrid reactor (60% and 38%). COD removal resulted from biological degradation and solids accumulation in the reactors. At 28°C, a constant 200 g sludge mass was observed in both reactors, and COD removal was attributed to biological degradation only. At lower temperatures, solids accumulation was observed in addition to biological degradation with an increase in reactor sludge as the temperature decreased. The decrease in biological degradation at lower temperatures was offset by solids accumulation and explains the similar overall COD removal efficiency observed at 28, 20, and 14°C. The decrease in temperature was also followed by an increase in the effluent TSS concentration in both reactors. At 14°C and 10°C, a lower effluent TSS concentration and better performance was observed in the UASB reactor (Lew et al. 2004).

The performance of an EGSB reactor coupled with hollow fiber membrane filtration for treating domestic wastewater was monitored during a seven-month period in the range 11–25°C, and at an HRT of 3.5 to 5.7 h. With temperatures above 15°C, the system was capable of removing 85–96% of total COD and 83–94% of TOC. At 11°C, increasing HRT from 3.5 to 5.7 h, the total COD removal was increased from 76 to 81%. Applying a higher up-flow velocity contributed to better effluent removal efficiency and a higher membrane permeability. The results of tests showed that cake layer resistance was the major resistance. More extracellular polymer substances (EPS) tended to be accumulated on the membrane surface than in the granules, since its content was higher on the membrane surface than in the granules (Chu, Yang, and Zhang 2005).

At low temperatures, the low hydrolysis rate and a decrease in the degradable organic matter fraction were found to cause the deterioration of the overall anaerobic reactor performance (Elmitwalli et al. 2001). UASB COD removals of ~65% at 20°C and of 55–65% at 13–17°C were observed by several authors (Lettinga et al. 1980; Elmitwalli et al. 1999; Seghezzo et al. 2000). A decrease in the effluent quality was also observed, together with a decline in the gas production rate. Agrawal et al. (1997) observed a 78% decrease in the gas production rate when the temperature was reduced from 27°C to 10°C. The low gas production coincided with a 25% lower COD removal at 10°C than at 27°C, indicating suspended solids accumulation in the reactor. One possible way to improve the performance of a UASB reactor at low temperatures is to provide surface area for biomass attachment and growth in the reactor volume above the sludge blanket (Tilche and Vieira 1991). This can be accomplished by replacing the typical gas/solids separator of the classical UASB reactor with filter media. Elmitwalli et al. (1999) compared the performances of a hybrid UASB filter and a classical UASB reactor at 13°C. The hybrid UASB filter reactor reached 64% COD removal, a 4% better removal than the classical UASB. A better colloidal fraction removal was attributed to the attached biomass on the filter (Lew et al. 2004).

The treatment of sewage at a temperature of 13°C was investigated in three reactors (each 3.84 L): a UASB and two AH reactors with small sludge granules with an average diameter of 0.73 mm. The media used in the AH reactors were vertical polyurethane foam sheets. The reactors were operated at an HRT of 8 h. The use of small sludge granules and operating the reactors at low up-flow velocity (1.8 m/d) improved suspended COD removal efficiencies for the UASB reactor. Moreover, the use of sheets in the AH reactors significantly increased suspended COD removal efficiencies as compared to the UASB and reached to 87% for pre-settled sewage treatment. The treatment of pre-settled sewage instead of raw sewage in AH reactors significantly increased colloidal and dissolved COD removal efficiencies with 13% and 12%, respectively, and colloidal COD removal efficiency for the UASB reactor with 13%. At "steady state" for presettled sewage treatment, the AH reactors removed 64% of the total COD, which is significantly higher, by 4%, than the UASB reactor. Therefore, the anaerobic treatment of domestic sewage at low temperatures can be improved by treating pre-settled sewage in shallow AH reactors containing small sludge granules (Elmitwalli et al. 1999).

4.4 ROLE OF EXTREMOPHILIC ANAEROBES

Extremophilic microorganisms are organisms capable of proliferating under extreme conditions that are generally detrimental to most life on Earth. Extremophilic anaerobes are widespread in saline, acid, alkaline, and high- or low-temperature environments. They are organisms considered of importance in different areas of research, due to their ability to produce proteins and enzymes under inhospitable conditions. Recent studies (González and y Terrón 2021; Hernández-Montoya et al. 2012) on the metabolic pathways of the extremophilic microorganisms and their environment have shown that most of the archaea, and some bacteria and cyanobacteria, carry out metabolic activities essential for the biogeochemical cycles of sulfur, carbon, and nitrogen. Archaea and bacteria being two of the main types of microorganisms that participate in a variety of processes, such as sulfidogenesis, methanogenesis, ANAMMOX (anaerobic ammonium oxidation), among others. This has suggested that extremophilic microorganisms and extreme ecosystems have a significant impact on global biogeochemical cycles (González and y Terrón 2021).

The ability of microorganisms to survive under harsh conditions has recently prompted researchers to study these microbes to better understand their characteristics and eventually utilize them in various applications. Extremophilic microorganisms withstand one or more extreme condition constraints due to cold, heat, high acidity, high alkalinity, high salinity, and high-pressure levels. These constraints would destroy normal microorganisms. Extremophiles can be classified according to the environment in which they survive. Many extremophilic microorganisms possess properties suitable for biotechnological and commercial uses (Tango and Islam 2002).

Extremophilic anaerobes that produce hydrogen via anaerobic fermentation are promising organisms for the conversion of biomass and organic wastes to hydrogen in open (non-sterile) systems in which the extreme process conditions efficiently prevent the growth of H_2-consuming microorganisms. Furthermore, extremophilic microorganisms are viable candidates for processing harsh (hot, highly saline, acidic, or alkaline) organic waste streams to valuable products. Recent studies have mainly focused on thermophiles due to the high H_2 production yields achieved at high temperature, while investigation on H_2 production by other extremophilic organisms is still scarce (Kivistö et al. 2013).

Microorganisms from extreme environments are especially important because they have enzymes and proteins that can work properly in extreme environmental conditions, such as, extreme temperatures, pH, salinity, and pressure. These microorganisms can be used in different biotechnological applications, providing great momentum for biofuel production (Mukhtar and Aslam 2020). Extremophilic microorganisms, including thermophiles, psychrophiles, halophiles, alkaliphiles, and acidophiles, have the ability to produce biofuels, such as bioethanol, biobutanol, biodiesel, and biogas or methane, by using various starting materials, such as sugars, starch crops, plant seeds, lignocellulosic agricultural waste, and animal waste, under extreme environments.

With progress being made in bioinformatics and gene-editing tools, microorganisms such as *Saccharomyces cerevisiae, Escherichia coli, Clostridium thermocellum, Pyrobaculum calidifontis*, and *Thermococcus kodakarensis* have been genetically engineered to upscale biofuel production (Mukhtar and Aslam 2020).

4.5 INDUSTRIAL AND TECHNOLOGICAL ASPECTS

4.5.1 Full-Scale Experiences with Anaerobic Treatment of Industrial Wastewater

High-rate anaerobic treatment for industrial wastewater was first applied on a commercial scale in the sugar industry in the mid-1970s. Since that time, the technology has developed into a standard method of wastewater treatment for a wide variety of industries (Frankin 2001).

Anaerobic treatment has gained tremendous success over the past two decades for treatment of industrial effluents. Apparent advantages are to be found in low operating costs, compact construction, production of energy through biogas, low surplus sludge production, resulting in overall favorable economics. For anaerobic treatment to compete with alternative technologies such as aerobic or physico-chemical treatment, it has to be cost effective (in terms of investment and operating costs), reliable, and durable. Anaerobic treatment is an established technology for a wide variety of industrial applications (Frankin 2001).

A four-cell anaerobic baffled reactor (ABR) was operated for two years treating raw municipal wastewater at ambient water and air temperatures of 12–23°C and −10 to 35°C, respectively. The 1000 L pilot reactor operated at a 12 h hydraulic residence time and was located in the Headworks building of the Plum Creek Water Reclamation Authority. The average influent was TSS = 510 ± 400 mg/L, BOD_5 = 320 ± 80 mg/L, and the average removal of TSS and BOD_5 was 83 ± 10% and 47 ± 15%, respectively. The TSS and BOD removal exceeded that of conventional primary clarification, with no wasting of the settled solids over the two years and stoichiometric production of methane. The estimated energy content of the biogas produced per unit volume of wastewater treated averaged 0.45 kWh/m^3. The TSS and total COD removal in the first cell averaged 75 ± 15% and 43 ± 14%, respectively, but methane production was only 20% of the total observed for the full ABR. The performance of the ABR relative to the extent of solids hydrolysis and methane production can be varied by the number of cells and hydraulic residence time. The anaerobic baffled reactor is an energy-positive technology that can be used for biologically enhanced primary treatment of raw municipal wastewater in cold climates (Hahn and Figueroa 2015).

Scully, Collins, and O'Flaherty (2006) demonstrated the (1) feasibility of psychrophilic, or low-temperature, anaerobic digestion (PAD) of phenolic wastewaters at 10–15°C; (2) economic attractiveness of PAD for the treatment of phenol as measured by daily biogas yields, and (3) impact on bioreactor performance of phenol loading rates (PLRs) in excess of those previously documented (1.2 kg phenol m^{-3} d^{-1}). Two EGSB-based bioreactors, R1 and R2, were employed to mineralize a volatile fatty acid-based wastewater. R2 influent wastewater was supplemented with phenol at an initial concentration of 500 mg L^{-1} (PLR, 1 kg m^{-3} d^{-1}). Reactor performance was measured by COD removal efficiency, CH_4 composition of biogas, and phenol removal (R2 only). The applied PLR was increased to 2 kg m^{-3} d^{-1} on day 147 and phenol removal by day 415 was 99% efficient, with ≤4 mg L^{-1} present in R2 effluent. The operational temperature of R1 (control) and R2 was reduced by stepwise decrements from 15°C through to a final operating temperature of 9.5°C. COD removal efficiencies of 90% were recorded in both bioreactors at the conclusion of the trial (day 673), when the phenol concentration in R2 effluent was below 30 mg L^{-1} (Scully, Collins, and O'Flaherty 2006).

The effect of sulfate addition on the stability of, and microbial community behavior in, low-temperature anaerobic expanded granular sludge bed-based bioreactors was investigated at 15°C. Efficient bioreactor performance was observed, with COD removal efficiencies of >90%, and a mean SO_4^{2-} removal rate of 98.3%. *In situ* methanogenesis appeared unaffected at a COD:SO_4^{2-} influent ratio of 8:1, and subsequently of 3:1, and was impacted marginally only when the COD:SO_4^{2-} ratio was 1:2. Specific methanogenic activity assays indicated a complex set of interactions between sulfate-reducing bacteria (SRB), methanogens, and homoacetogenic bacteria. SO_4^{2-} addition resulted in predominantly acetoclastic, rather than hydrogenotrophic, methanogenesis until >600 days of SO_4^{2-} influenced bioreactor operation. Temporal microbial community development was monitored by denaturation gradient gel electrophoresis (DGGE) of 16S rRNA genes. Fluorescence *in situ* hybridizations (FISH), qPCR, and microsensor analysis were combined to investigate the distribution of microbial groups, and particularly SRB and methanogens, along the structure of granular biofilms. qPCR data indicated that sulfidogenic genes were present in methanogenic and sulfidogenic biofilms, indicating the potential for sulfate reduction even in bioreactors not exposed to SO_4^{2-}. Although the architecture of methanogenic and sulfidogenic granules was similar, indicating the presence of SRB even in methanogenic systems, FISH with rRNA targets found that the SRB were more abundant in the sulfidogenic biofilms. Methanosaeta species were the predominant, keystone members of the archaeal community, with the complete absence of the Methanosarcina species in the experimental bioreactor by trial conclusion. Microsensor data suggested the ordered distribution of sulfate reduction and sulfide accumulation, even in methanogenic granules (Madden et al. 2014).

EGSB bioreactors were used to investigate SRB activity in low-temperature anaerobic digesters (Madden et al. 2014). The impact of sulfate contamination on methanogenesis, as well as on community structure, and the distribution and abundance of SRB functional genes, was assessed at different COD:SO_4^{2-} ratios (Madden et al. 2014). COD removal can proceed at 15°C in anaerobic digesters exposed to sulfate. *In situ* methane production appears impacted only at COD:SO_4^{2-} ratios ≤1:2; thus, higher COD:SO_4^{2-} ratios would appear to support biogas production in cold anaerobic digesters. Hydrogenotrophic methanogens in low-temperature anaerobic sludge granules were more sensitive to sulfate than acetoclastic methanogens, but complex interactions of SRB, methanogens, and homoacetogenic bacteria appear to underpin COD removal by sulfate reduction and methanogenesis (Madden et al. 2014).

Two strategies exist for seeding low-temperature anaerobic reactors: the use of specialist psychrophilic biomass and mesophilic bioreactor sludge acclimated to low temperature (Bowen et al. 2014). These strategies determine the low-temperature limitation of anaerobic sludge from a bioreactor acclimated to United Kingdom (UK) temperatures (<15°C). Anaerobic incubation tests using low-strength DWW and various alternative soluble COD sources were conducted at 4, 8, and 15°C; methanogenesis and acidogenesis were monitored separately. Production of methane and acetate was observed; decreasing temperature resulted in decreased yields and increased "start-up" times. At 4°C methanogenesis, not hydrolysis/acidogenesis, was rate limiting. The final methane yields at 4°C were less than 35% of the theoretical potential, whereas at 8°C and 15°C more than 75% and 100% of the theoretical yield was achieved, respectively. It proposed that the lower temperature limit for DWW treatment with anaerobic bioreactor sludge was between 8°C and 4°C and that 8°C was the threshold for reliable operation (Bowen et al. 2014).

The feasibility of long-term (more than three years), low-temperature (4–15°C), and anaerobic bioreactor operation, for the treatment of acidified wastewater, was investigated (Lettinga et al. 1999). A hybrid, expanded granular sludge bed–anaerobic filter bioreactor was seeded with a mesophilic inoculum and employed for the mineralization of moderate-strength (3.75–10 kg COD m^{-3}) volatile fatty acid-based wastewaters at 4–15°C. Despite transient disimprovement, COD removal efficiency (CODRE) and methane biogas concentrations exceeded 80% and 65%, respectively, at an applied organic loading rate (OLR) of 10 kg COD m^{-3} d^{-1} between 9.5°C and 15°C (sludge loading rate (SLR), 0.6 kg COD kg[VSS]$^{-1}$ d^{-1}). More than 50% of the granular sludge bed was lost to disintegration during operation at 9.5°C, warranting a reduction in the applied OLR to 3.75–5 kg COD m^{-3} d^{-1} (SLR, ~0.4–0.5 kg COD kg[VSS]$^{-1}$ d^{-1}). From that point forward, remarkably stable and efficient performance was observed during operation at 4–10°C, with respect to CODRE (≥82%), methane biogas concentration (>70%), and methane yields (>4 $l_{Methane}$ d^{-1}), suggesting the adaptation of our mesophilic inoculum to psychrophilic operating conditions.

The data suggested that mesophilic inocula could physiologically adapt to sub-optimal operational temperatures; treatment efficiencies and sludge loading rates at 4°C (day 1243) were comparable to those achieved at 15°C (day 0). Furthermore, long-term, low-temperature bioreactor operation may act as a selective enrichment for psychrophilic methanogenic activity from mesophilic inocula. The observed efficient and stable bioprocess performance highlights the potential for long-term, low-temperature bioreactor operation (McKeown et al. 2009).

Anaerobic treatment of a volatile fatty acid (VFA) mixture was investigated under psychrophilic (3–8°C) conditions in two laboratory-scale EGSB reactor stages in series. The reactor system was seeded with mesophilic methanogenic granular sludge and fed with a mixture of VFAs. Good removal of fatty acids was achieved in the two-stage system. Relatively high levels of propionate were present in the effluent of the first stage, but propionate was efficiently removed in the second stage, where a low hydrogen partial pressure and a low acetate concentration were advantageous for propionate oxidation. The specific VFA-degrading activities of the sludge in each of the modules doubled during system operation for 150 days, indicating a good enrichment of methanogens and proton-reducing acetogenic bacteria at such low temperatures. The specific degradation rates of butyrate, propionate, and the VFA mixture amounted to 0.139, 0.110, and 0.214 g of chemical oxygen demand g of volatile suspended solids^{-1} day^{-1}, respectively. The biomass which was obtained after 1.5 years still had a temperature optimum of between 30°C and 40°C (Lettinga et al. 1999).

It indicated that a high-rate anaerobic treatment in a two-stage EGSB system was feasible under very-low-temperature conditions (i.e., down to 3°C). For a VFA mixture as the substrate, COD removal efficiencies of 90% and higher can be achieved at 8°C and 4°C at OLRs of 12 g and 5 g of COD liter^{-1} day^{-1}, respectively. In the two-module reactor system, methanogenic communities were obtained that degraded VFAs, including propionate, completely to CH_4 and CO_2 at low temperature. A psychrophilic population of microorganisms was not obtained. However, the EGSB reactor configuration enabled a high rate of conversion at a low temperature (Lettinga et al. 1999).

The anaerobic biological treatment of sucrose-based, low-strength wastewater was investigated in EGSB reactors at low temperatures over a 300-day trial period. During the trial, the operating temperature was lowered in a stepwise manner from 20°C to 5°C. As a result, the reactors exhibited sufficient performances until 10°C operation. The COD removal rate was 3.1–3.8 kg COD m^{-3} day^{-1} at 10°C. In particular, the COD removal rate increased gradually through the low-temperature operation; indeed, the later stages of the 10°C operation attained a rate similar to those achieved at 20°C and 15°C. This finding is especially practical for applications of psychrophilic methane fermentation. Additionally, the structure of the microbial community in the granular sludge was analyzed by clone analysis based on 16S rRNA genes and fluorescence *in situ* hybridization (FISH). As a result, the percentage of the phylum Firmicutes, which were assumed to be *Anaerobivrio* sp. and *Lactococcus* sp., greatly increased, from 0.7% to 8.0% of the total cells, especially in the surface layer of the granular sludge. These bacteria would contribute to the degradation of the sucrose substrate anaerobically at ambient temperatures. Moreover, the results suggest that a *Methanospirillum* species, which is a H_2-utilizing methanogen, increased from 0.5% to 6.7% during the low-temperature incubation, with a significant increase of methanogenic activity from H_2/CO_2 at 20°C. Thus, the *Methanospirillum* species detected in this study may have a key role as a hydrogen scavenger during hydrogen metabolism in low-temperature conditions (Tsushima et al. 2010).

4.5.2 Technological Aspects

The treatment of decentralized low-carbon greywater in rural areas, particularly in cold weather, remains a challenge. Oxic/anoxic (O/A) process and Fe/C micro-electrolysis were incorporated into vertical constructed wetland to develop "micro-electrolysis-oxic/anoxic constructed wetland" (ME-(O/A)CW) for practical decentralized low-carbon greywater treatment. ME-(O/A)CW provided NH_4^+-N, TN, TP, and COD removal of 94.3%, 86.2%, 98.0%, and 92.7%, respectively, at a hydraulic loading rate of 0.9 $m^3/(m^2·d)$ under low ambient temperature of −11.5°C to 8.0°C. Effective nitrification, phosphorus accumulating, and organic degradation were proceeded in the aerobic layers, and efficient H_2-/Fe^{2+}-mediated autotrophic denitrification and Fe^{3+}-based phosphorus immobilization were developed in the anaerobic layers through *in situ* H_2-/Fe^{2+} supply by Fe/C micro-electrolysis. Ammonium-oxidizing bacteria (AOB) (e.g. *Nitrosomonadales*), nitrite-oxidizing bacteria (NOB), phosphorus-accumulating organisms (PAOs) (e.g. *Nitrospira*), autotrophic denitrificans (e.g. *Thiobacillus*, *Hydrogenophaga*, and *Sulfurimonas*), heterotrophic denitrificans (e.g. *Denitratisoma*), and Fe(II)-oxidizing bacteria (e.g. *Ferritrophicum*) dominated ME-(O/A)CW and confirmed the reaction mechanisms. The developed ME-(O/A)CW presented significant potential in the practical application for decentralized low-carbon greywater treatment under low ambient temperature (Deng et al. 2020).

Although most of the high-rate reactors have proved their applicability for different high-strength wastewaters over a range of organic loading rates, certain differences exist in the preference of a particular type of digester over others in terms of various factors such as requirement of pretreatment, dilution, control of operating conditions, etc. In the case of slaughterhouse wastewater, an anaerobic contact reactor can be used without pretreatment, whereas for the usage of high-rate digesters such as UASB, a pretreatment step for removal of the suspended solids and fats is essential prior to anaerobic treatment. Two-phase digestion with pH and temperature control results in a higher biogas production rate with cheese whey wastewater digestion. An aerobic post-treatment is necessary to attain the permissible COD and BOD levels before discharge. Due to the generation of wastewater from various sections of the pulp and paper industry, there are variations in the composition and the treatability of effluents. Hence, it is preferable to treat the effluents from each section separately, depending on their biodegradability and suitability to the digestion process rather than treating the combined effluent. Advanced methods, such as coupling of reactors for suitable pretreatment and post-treatment, can result in complete treatment of the effluents with the acceptable limits. It is clear from the review that there are no governing factors that dictate the suitability of any particular reactor design for a specific effluent. By suitable modifications in the reactor designs and by altering the effluent characteristics, the existing high-rate digesters can be accommodated for treatment of organic effluents. However, based on the characteristics of the different reactors, such as efficiency based on loading rate and COD reduction, biomass retention, and other factors like cost, operation, and maintenance requirements, UASB and fixed film configurations appear to be the most suitable (Saleh and Mahmood 2004).

4.5.3 A Few Examples and Case Studies of Decentralized Anaerobic WWT in Cold Climates

Luostarinen et al. (2007) investigated the impact of seasonal temperature changes and low temperature in northern Europe on the effectiveness of UASB septic tanks in treating blackwater. Three UASB septic tanks were monitored under different operational parameters and temperatures. The study found that the UASB septic tank was a feasible method for (pre)treatment of blackwater at low temperatures, as it was effective in removing suspended solids and dissolved organic material. However, the inoculum sludge had a limited impact on CODss removal. In the start-up phase, some poorly adapted inoculum was disintegrated and washed out, which should have been taken into consideration while designing the process. Removal of CODdis was initially negative but improved as the sludge adapted to low temperature. Nevertheless, the UASB septic tank alone did not meet the Finnish or Dutch treatment requirements, and it should be considered mainly as a pretreatment method. Measuring the requirements as mgCOD l^{-1} may not always be the best method, as the volume of effluent discharged is also an essential factor that affects the final amount of COD entering the receiving water bodies (Luostarinen et al. 2007).

A pilot-scale UASB-Settler-Digester (USD) system (Xu et al. 2018) was employed to treat raw municipal wastewater obtained from a sewer system at a temperature of 10°C. During the reactor operation, UASB sludge was continuously transferred from the UASB to a settler, where the concentrated sludge was then transferred to a digester operated at 35°C. The study revealed that the settler, with a hydraulic retention time (HRT) of 3 h, increased UASB sludge chemical oxygen demand (COD) concentration from 14.5 ± 2.5 g/L to 29.9 ± 4.1 g/L. With an HRT of 6 h, the USD system achieved a mean COD removal efficiency of 49.2%, and 23.9% of influent COD was converted to methane. The specific methanogenic activities at 35°C of the UASB and the digester sludge were 0.26 g and 0.24 g CH_4 COD/(g VSS d), respectively, and the stability values were 0.21 g and 0.16 g CH_4 COD/g COD, respectively. The stability of the settled sludge was similar to that of the recirculated UASB sludge. In comparison to a UASB-digester system, the USD system with an added settler achieved similar COD removal and methane production, but with a reduced sludge recirculation rate (from 16% to 8% of the influent flow rate), resulting in a 50% reduction in heating energy consumption in the digester of the UASB digester (Xu et al. 2018).

These examples demonstrate that decentralized anaerobic WWT systems can be effective in cold climates with proper design and operation considerations.

4.6 TECHNO-ECONOMIC FEASIBILITY

The existing wastewater treatment infrastructure has not adequately established an efficient and sustainable use of energy, water, and nutrients. A proposed scheme based on source separation and water-efficient use is compared to the current wastewater management paradigm (one largely based on activated sludge) using techno-economic terms. Garrido-Baserba et al. (2018) explored the economic viability of adopting more sustainable management alternatives and expanded the understanding of the economics of decentralization and source separation. The feasibility of three different potential types of source separation (with different levels of decentralization) are compared to the conventional centralized activated sludge process by using recognized economic assessment methodologies together with widely accepted modeling tools. The alternatives were evaluated for two common scenarios: new developments and retrofit due to the aging of existing infrastructures. The results prove that source-separated alternatives can be competitive options despite existing drawbacks (only when countable incomes are included), while the hybrid approach resulted in the least cost-effective solution. A detailed techno-economic evaluation of the costs of decentralization provides insight into the current constraints concerning the paradigm shift and the cost of existing technologic inertia (Garrido-Baserba et al. 2018).

To provide food waste energy recovery, sludge management, and building wastewater recycling with the primary goal of achieving zero-waste discharge, a novel integrated single-stage anaerobic co-digestion and moving bed biofilm reaction membrane bioreactor system has been developed. The acceptability of the technology and the sustainability of the zero-waste system were assessed by a proposed stepwise evaluation approach that included the factors affecting its usage, technical performance, environmental sustainability, and economic feasibility. Based on the three experimental cases of the system prototype, the anaerobic co-digestions of food waste and wastewater sludge (Cases 2 and 3) provided slightly higher methane generation rates than were achieved from the mono-digestion of food waste (Case 1). Average methane yields from the anaerobic co-digestion (Cases 2 and 3) were in the range of 0.352–0.353 L/g TVS. The ratios of VFA/Alk of all the experimental cases were lower than 0.25, indicating the stable anaerobic co-digestion performances. The MBBR-MBR was able to achieve an ammonia removal rate in the range of 0.34–0.77 g NH_4/m^2-day. The co-treatment of combined wastewater was able to provide good reclaimed water that was suitable for landscape use. Life cycle assessment results showed that the proposed zero-waste system could reduce the impacts of climate change, eutrophication, acidification, and fossil resource depletion compared to conventional waste treatment. The economic Internal Rate of Return, Net Present Value, and Benefit/Cost of the system were around 9%, 2222 USD, and 1.44, respectively. The challenges and opportunities from implementing the system were discussed (Sakcharoen, Ratanatamskul, and Chandrachai 2021).

This study developed an integrated system of anaerobic co-digestion and a moving bed biofilm reaction-membrane bioreactor for food waste energy recovery, sludge management, and building wastewater recycling with the primary goal of achieving zero-waste discharge. The results revealed that the essential factors influencing the intention to use a zero-waste system are water recycling efficiency, management attitude, biogas production efficiency, and effluent quality after treatment (Sakcharoen, Ratanatamskul, and Chandrachai 2021).

The up-flow anaerobic hybrid reactor (UAHR) was simultaneously employed for two-biofuels (H_2 and CH_4) production from solely biscuits wastewater (BW) industry and a mixture of BW with domestic sewage (DS) at a mixing ratio of 3:1. The UAHR was fed with BW and operated at an organic loading rate (OLR) of 21.9 ± 1.7 gCOD/L/d, and hydraulic retention time (HRT) of 12.0 h. The module achieved volumetric hydrogen production rate (VHPR) and hydrogen yield (HY) of 0.95 ± 0.13 L/L reactor/d, and 118.6 ± 22.3 mL/gCOD$_{removed}$, respectively. A substantial deterioration in the H_2 productivity occurred after 94 days of continuous-feed operation, due to the inhibition effect of long-chain fatty acids (LCFAs). DS was integrated into the BW at a mixing ratio of 3.0 (BW):1.0 (DS) where the UAHR was continuously operated at OLRs of 7.95 and 4.45 gCOD/L/d, achieving H_2 and CH_4 yields of 414.2 ± 38.1 and 318.5 ± 23.5 mL/gCOD$_{removed}$, respectively. The payback periods for the production of biogas-rich CH_4 and H_2 were 7.1 and 5.7 y, respectively. The anaerobic digestion of BW industry was investigated. H_2 yield of 118.6 mL/gCOD removed was obtained by solely anaerobic digestion of BW. OLR of 4.45 gCOD/L/d provided MY of 318.5 ± 23.5 mL/g CODt removed. HY of 414.2 ± 38.1 mL/g CODt removed was achieved at OLR exceeding 7.95 gCOD/L/d. Techno-economic study of BW treatment showed suitable payback period for investment (Tawfik et al. 2021).

Decentralized treatment is often assumed to have the potential to improve the technical, environmental, and economic performance of current technologies. The techno-economic feasibility of implementing independent building-scale decentralized systems combining rainwater harvesting, potable water production, and wastewater treatment and recycling was assessed for six main types of buildings ranging from single-family dwellings to high-rise buildings. Five different treatment layouts were evaluated under five different climatic conditions for each type of building. The layouts considered varying levels of source separation (i.e. black, grey, yellow, brown, and combined wastewater) using the corresponding toilet types (vacuum, urine-diverting, and conventional) and the appropriate pipes and pumping requirements. The results indicated that the proposed layouts could satisfy 100% of the water demand for the three smallest buildings in all but the most arid climate conditions. For the three larger buildings, rainwater would offset annual water needs by approximately 74–100%. A comprehensive economic analysis considering capital expenditures (CapEx) and operational expenditures (OpEx) indicated that the cost of installing on-site water harvesting and recycling systems would increase the overall construction cost of multifamily buildings by around 6% and single-family dwellings by about 12%, with relatively low space requirements. For buildings or combined water systems with more than 300 people, the estimated total price of on-site water provision (including harvesting, treatment, recycling, and monitoring) ranged from \$1.5/m^3 to \$2.7/m^3, which is considerably less than the typical tariffs collected by utilities in the United States and Western Europe. Where buildings can avoid the need to connect to centralized supplies for potable water and sewage disposal, water costs could be even lower. Urine-diversion has the potential to yield the least expensive solution, but it is the least well developed and had higher uncertainty in the cost analysis. More mature layouts (e.g. membrane bioreactors) exhibited less cost uncertainty and were economically competitive. Our analysis indicates that existing technologies can be used to create economically viable systems that greatly reduce demands on centralized utilities and, under some conditions, eliminate the need for centralized water supply or sewage collection (Garrido-Baserba et al. 2022).

Nowadays, several technologies and management strategies are proposed for upgrading wastewater treatment plants (WWTPs) in order to improve sludge stabilization or minimize sludge production. In order to make sure that this leads to overall improvements, their techno-economic and environmental performance has to be carefully evaluated. In response to this, the techno-economic and environmental performance was assessed for a considered model reference WWTP upgraded with separate treatment and disposal for primary and secondary sludge to provide guidance to research, policy, and investment efforts. It departs in an experimental campaign carried out within the EU project "ROUTES" and describes an assessment of an upgrade of a WWTP, involving wet oxidation of primary sludge and thermophilic anaerobic digestion (integrated with thermal hydrolysis pretreatment) for secondary sludge. The reference plant and the upgrade implied different disposal routes for the resulting sludges: incineration for the stabilized sludge deriving from the reference WWTP; agricultural land application for the secondary stabilized sludge from the upgraded plant and landfilling for the solid residue deriving from the wet oxidation process (as nonhazardous waste).

The mass balance carried out for both reference and upgraded WWTPs showed a reduction of total sludge production (up to 45%) for the upgraded WWTP. The additional energy demand of the upgraded solution (+22.4 Wh/(PE d), PE: population equivalent) was partially balanced by the electric energy produced by combined heat and power unit installed in the upgraded WWTP (–18.08 Wh/(PE d)). The integrated assessment showed that the upgraded solution, although resulting in an increase of WWTP technical complexity, may contribute to environmental and economic advantages. The paper provides guidance in terms of which aspects need a more thorough evaluation when this upgrade is considered in real cases (Briassoulis, Pikasi, and Hiskakis 2021).

Estimations of running costs revealed that anaerobic treatment of sewage at less than 20°C was more economical than aerobic treatment. An anaerobic fluidized-bed reactor (AFBR) and a UASB were applied to the anaerobic treatment of sewage at 20°C. Even at an HRT of 6 h, in both cases the quality of the effluent was 5 mg/L of BOD with a TOC removal efficiency of 80%, which would be appropriate for discharge into rivers. However, at an HRT of 3 h, TOC removal efficiency decreased to 70%, while in an activated sludge process (21°C, HRT 3 h), the TOC removal efficiency was maintained at 83.8%. In order to investigate the effect of temperature on treatment, the temperature was decreased to 13°C in stepwise fashion at an HRT of 6 h. In both the AFBR and UASB, temperatures above 15°C did not affect the TOC removal efficiency, but the TOC concentration in the effluent at 13°C increased to 12.5 mg/L in the AFBR and to 15 mg/L in the UASB. However, at an HRT of 6 h and temperatures above 15°C as the standard conditions, the quality of the effluent treated anaerobically by the AFBR or UASB was almost the same as that by the activated sludge process (21°C, HRT 6 h) (Kida, Tanemura, and Sonoda 1993).

The studies emphasized the importance of techno-economic feasibility in evaluating the practicality of proposed wastewater treatment solutions. Although there were some limitations, the studies demonstrated that source separation and water-efficient use, anaerobic co-digestion and moving bed biofilm reaction-membrane bioreactor, and up-flow anaerobic hybrid reactors were all financially feasible and cost-effective options for wastewater treatment. These findings underscore the significance of considering both technological and economic aspects when assessing potential solutions for wastewater management.

4.7 CHALLENGES IN COLD CONDITIONS

Temperature is the bottleneck for the anaerobic treatment of domestic wastewater in temperate climates. Most previous attempts to achieve anaerobic treatment at low temperatures have attempted to acclimatize mesophilic sludge and have failed at temperatures below 10–13°C.

At low temperatures, the low hydrolysis rate and a decrease in the degradable organic matter fraction were found to cause the deterioration of the overall anaerobic reactor performance, with temperature decrease, the rate of substrate utilization may be reduced.

While temperature decrease, maximum microbial specific growth may reduce; at lower temperature, methane may become more soluble in the reactor liquor (McKeown et al. 2012).

In tropical climates with higher ambient temperatures, intensive anaerobic systems such as the UASB reactor have been successfully applied and found wide acceptance for domestic wastewater treatment. COD removals above 70% have been reported by several authors (Kalogo and Verstraete 2000; Lew et al. 2004). At lower ambient temperatures, overall performance deteriorated, with COD removals of ~65% at 20°C and 55–65% at 13–17°C (Al-Jamal and Mahmoud 2009; Lew et al. 2004).

In light of the growing energy crisis, the study of AnMBR for domestic wastewater treatment at low temperatures is especially important, particularly for temperate and cold regions. Smith, Skerlos, and Raskin (2015) revealed that the COD removal rate of an anaerobic membrane bioreactor could surpass 95% at 6°C but declined to 86% at 3°C. The decrease in the dissolved COD content in the sludge mixture indicated a reduction in microbial activity. The high COD removal rate at low temperatures in the reactor was attributed to the biofilm effect on the surface of the membrane module. Although the start-up period of the anaerobic membrane bioreactor at low temperatures is lengthier and the treatment effect and gas production are lower than at medium temperature conditions (33–35°C), the electrical energy consumption required to operate the anaerobic membrane bioreactor is comparatively lower.

Therefore, additional research is needed to address temperature and energy consumption issues, to promote the widespread applications of AnMBRs. The adverse effects of low temperatures on the AnMBR can be compensated for by extending the hydraulic residence time, but also by increasing the operating costs if the residence time is too long. To solve these problems, the shortcomings of low-temperature operation can be compensated for by developing new membrane modules, improving start-up operation at low temperatures and using the biogas produced for reactor warming, low temperatures and an increase in organic loading can promote the release of EPS, resulting in faster membrane contamination rates.

Another crucial aspect that warrants investigation is the efficient and cost-effective recovery of dissolved methane using AnMBRs (anaerobic membrane bioreactors). This undertaking necessitates achieving optimal methane production while minimizing its dissolution at low temperatures (10–30°C). It entails developing methane recovery technology, designing appropriate methane recovery equipment, enhancing cost management, and ensuring energy consumption efficiency throughout the recovery process.

To improve the performance of a UASB reactor treating raw domestic wastewater under temperate climates conditions, the addition of a sludge digester to the process was investigated. With the decrease in temperature, the COD removal decreased from 78% at 28°C to 42% at 10°C for the UASB reactor operating alone at a hydraulic retention time of 6 h. The decrease was attributed to low hydrolytic activity at lower temperatures that reduced suspended matter degradation and resulted in solids accumulation in the top of the sludge blanket. Solids removed from the upper part of the UASB sludge were treated in an anaerobic digester. Based on sludge degradation kinetics at 30°C, a digester of 0.66 L per liter of UASB reactor was designed to operate at a retention time of 3.20 days. Methane produced by the sludge digester is sufficient to maintain the temperature at 30°C (Lew et al. 2011).

At low temperatures, the effectiveness of intensive anaerobic systems is significantly hindered by the hydrolysis of suspended solids and/or the reduction in organic matter degradation, leading to their accumulation in the sludge bed (Al-Jamal and Mahmoud 2009; Lew et al. 2004). It has been observed that the relatively high concentration of particulate matter present in domestic wastewater (approximately 350 mg/L total suspended solids) exhibits lower degradation rates under psychrophilic temperatures. Consequently, lower temperatures result in an increased production of excess sludge, which reduces the sludge retention time (SRT). A shorter SRT can impede the growth of methanogenic microorganisms, leading to inadequate removal of soluble COD and a deterioration in sludge stability. Thus, the efficiency of anaerobic processes is heavily reliant on the local ambient temperatures.

Anaerobic digestion in cold climates for domestic, house on-site, biogas generation is hardly practiced worldwide. The prime reasons for the limited adoption are the capital investments required to overcome the impact of the low temperatures on biogas production. In relatively mild climatic winters, with average ambient temperatures of 5–10°C, relatively simple measures are adequate to avoid a decrease in biogas production. However, in countries with more severe winters, active heating is necessary. One of the possibilities studied is to capture solar energy for digester heating using solar collectors.

The challenge now is not only to build such reactors and determine their ability to attain and retain a satisfactory abundance of adapted bacteria and methanogens under realistic conditions but to demonstrate that such reactors would indeed have a positive contribution to a carbon-neutral energy-positive wastewater treatment industry (Smith et al. 2014).

At low temperatures, the viscosity of liquids is also increased. Therefore, more energy is required for mixing, and sludge bed reactors become less easily mixed, particularly at low biogas production rates. In psychrophilic reactors, particles will settle slower because of a decreased liquid–solid separation at low temperatures. Moreover, related to the higher liquid viscosity, the diffusion of soluble compounds will drop at lower temperatures. The diffusion constant of soluble compounds is about 50% lower at 10°C compared with the mesophilic temperature range (30–40°C) (Lettinga et al. 2001).

Anaerobic treatment of domestic wastewater can also be very interesting and cost-effective in countries where the priority in discharge control is removal of organic pollutants. The economy and technologies today largely depend upon energy resources that are not renewable. It is therefore necessary to identify and develop alternative sources of energy that are sustainable (Mrowiec and Suschka 2009).

4.8 CONCLUSIONS AND PERSPECTIVES

Anaerobic wastewater treatment systems hold immense promise for addressing wastewater challenges in cold environments. Despite being influenced by temperature, pH, hydraulic retention time (HRT), and other factors, these systems boast numerous advantages, including energy efficiency, high removal of COD, methane production, and low sludge yield. Moreover, their potential to align with treatment guidelines through innovative combined treatment configurations enhances their appeal and usefulness in real-world applications.

The advantages offered by anaerobic treatment in cold environments are significant. The ability to achieve high COD removal rates and produce methane gas, a valuable source of renewable energy, underscores the environmental and economic benefits of this approach. Additionally, the reduced sludge production decreases the burden on waste disposal, making anaerobic systems a greener and more sustainable alternative.

However, certain challenges remain in the successful implementation of anaerobic treatment in cold climates. The sensitivity of these systems to temperature fluctuations and other environmental factors necessitates ongoing research and development to optimize their performance. Specifically, efforts should be directed toward creating new and advanced configurations that can increase energy efficiency, allowing these systems to reach commercial or industrial scales. By enhancing the technology's robustness and reliability, its widespread adoption becomes more viable, benefiting communities and industries alike.

Investing in research and development for anaerobic wastewater treatment in cold environments is vital. By continually exploring new innovations, the strategy can mature into a more cost-effective and practical solution for wastewater management. Governments, institutions, and private sectors must allocate resources and funding to support such research initiatives, recognizing the potential long-term benefits it can bring in terms of environmental preservation and energy sustainability.

Collaboration among researchers, engineers, and policymakers is key to overcoming challenges and propelling anaerobic treatment into mainstream use. Knowledge sharing and communication among stakeholders can foster a conducive environment for technology adoption, facilitate the exchange of best practices, and accelerate the integration of anaerobic systems into wastewater treatment practices.

The future of wastewater treatment lies in the seamless integration and optimization of diverse technologies, and anaerobic systems have proven to be a valuable addition to this landscape. As advancements continue and the full potential of anaerobic treatment in cold environments is unlocked, we can anticipate a remarkable transformation in wastewater treatment approaches. This transformation will be characterized by environmentally friendly, cost-effective, and sustainable solutions that benefit communities, industries, and the planet as a whole.

Anaerobic wastewater treatment in cold environments holds great promise for addressing critical water management challenges. While there are obstacles to overcome, continuous research and development efforts will undoubtedly lead to improved technology, making it an even more appealing choice for decentralized wastewater treatment. By fostering collaboration and investment in this field, it can pave the way for a future where wastewater is efficiently and sustainably managed, contributing to a cleaner and healthier world for generations to come.

REFERENCES

Agrawal, K.R., Lucas, P.W., Prinz, J.F., and Bruce, I.C. (1997). "Mechanical properties of foods responsible for resisting food breakdown in the human mouth." *Archives of Oral Biology* 42(1): 1–9. ISSN 0003-9969, https://doi.org/10.1016/S0003-9969(96)00102-1.

Al-Jamal, W., and Mahmoud, N. (2009). "Community onsite treatment of cold strong sewage in a UASB-septic tank." *Bioresource Technology* 100(3): 1061–1068.

Bashiri, Reihaneh, Ben Allen, Burhan Shamurad, Martin Pabst, Thomas P Curtis, and Irina D Ofiţeru. 2022. "Looking for lipases and lipolytic organisms in low-temperature anaerobic reactors treating domestic wastewater." *Water Research* 212:118115.

Bowen, Emma J, Jan Dolfing, Russell J Davenport, Fiona L Read, and Thomas P Curtis. 2014. "Low-temperature limitation of bioreactor sludge in anaerobic treatment of domestic wastewater." *Water Science and Technology* 69 (5):1004–1013.

Briassoulis, Demetres, Anastasia Pikasi, and Miltiadis Hiskakis. 2021. "Organic recycling of post-consumer/industrial bio-based plastics through industrial aerobic composting and anaerobic digestion - techno-economic sustainability criteria and indicators." *Polymer Degradation and Stability* 190:109642. doi: https://doi.org/10.1016/j.polymdegradstab.2021.109642.

Chu, Li-Bing, Feng-Lin Yang, and Xing-Wen Zhang. 2005. "Anaerobic treatment of domestic wastewater in a membrane-coupled expended granular sludge bed (EGSB) reactor under moderate to low temperature." *Process Biochemistry* 40 (3-4):1063–1070.

Deng, Shihai, Binghan Xie, Qiang Kong, Shuai Peng, Hengchen Wang, Zhifeng Hu, and Desheng Li. 2020. "An oxic/anoxic-integrated and Fe/C micro-electrolysis-mediated vertical constructed wetland for decentralized low-carbon greywater treatment." *Bioresource Technology* 315:123802. doi: https://doi.org/10.1016/j.biortech.2020.123802.

Elmitwalli, Tarek A, Marcel H Zandvoort, Grietje Zeeman, Harry Bruning, and Gatze Lettinga. 1999. "Low temperature treatment of domestic sewage in upflow anaerobic sludge blanket and anaerobic hybrid reactors." *Water Science and Technology* 39 (5):177–185.

Elmitwalli, T. A., S. Sayed, L. Groendijk, J. van Lier, G. Zeeman, and G. Lettinga. 2003. "Decentralised treatment of concentrated sewage at low temperature in a two-step anaerobic system: Two upflow-hybrid septic tanks." *Water Science and Technology* 48 (6):219–226. doi: https://doi.org/10.2166/wst.2003.0404.

Elmitwalli, T. A., Soellner, J., De Keizer, A., Bruning, H., Zeeman, G., and Lettinga, G. 2001. "Biodegradability and change of physical characteristics of particles during anaerobic digestion of domestic sewage." *Water Research* 35(5): 1311–1317.

Frankin, R. J. 2001. "Full-scale experiences with anaerobic treatment of industrial wastewater." *Water Science and Technology* 44 (8):1–6.

Garrido-Baserba, Manel, Irene Barnosell, Maria Molinos-Senante, David L. Sedlak, Korneel Rabaey, Oliver Schraa, Marta Verdaguer, Diego Rosso, and Manel Poch. 2022. "The third route: A techno-economic evaluation of extreme water and wastewater decentralization." *Water Research* 218:118408. doi: https://doi.org/10.1016/j.watres.2022.118408.

Garrido-Baserba M, Vinardell S, Molinos-Senante M, Rosso D, Poch M. 2018. "The economics of wastewater treatment decentralization: A techno-economic evaluation." *Environ Sci Technol* 52(15):8965–8976.

Gavala, Hariklia N, Irini Angelidaki, and Birgitte K Ahring. 2003. "Kinetics and modeling of anaerobic digestion process." *Biomethanation I*:57–93.

González, Arlette Galván, and Rocío Pérez y Terrón. 2021. "Importance of extremophilic microorganisms in biogeochemical cycles." *GSC Advanced Research and Reviews* 9 (1):082–093.

González, A., and Pérez Y Terrón, R. 2021. "Importance of extremophilic microorganisms in biogeochemical cycles. " *GSC Advanced Research and Reviews*, 9: 082–093.

Hahn, Martha J., and Linda A. Figueroa. 2015. "Pilot scale application of anaerobic baffled reactor for biologically enhanced primary treatment of raw municipal wastewater." *Water Research* 87:494–502. doi: https://doi.org/10.1016/j.watres.2015.09.027.

Halalsheh, M.M. 2002 *"Anaerobic pre-treatment of strong sewage: a proper solution for Jordan."* Thesis, Wageningen University, the Netherlands.

Hejnic, Jakub, Petr Dolejs, Vojtech Kouba, Andrea Prudilova, Patria Widiayuningrum, and Jan Bartacek. 2016. "Anaerobic treatment of wastewater in colder climates using UASB reactor and anaerobic membrane bioreactor." *Environmental Engineering Science* 33 (11):918–928. doi: https://doi.org/10.1089/ees.2016.0163.

Hernández-Montoya V, Alvarez LH, Montes-Morán MA, Cervantes FJ 2012. Reduction of quinone and non-quinone redox functional groups in different humic acid samples by Geobacter sulfurreducens. *Geoderma* 183:25–31. https://doi.org/10.1016/j.geoderma.2012.03.007.

Hulshoff-Pol, LW. 1998. "Waste characteristics and factors affecting reactor performance." Lecture notes by *H*ulshoff *P*ol in *I*nternational *C*ourse on *A*naerobic *W*astewater *T*reatment, Wageningen Agricultural University, *The Delft,* Netherlands.

Hussain, A., and Dubey, S. K. (2017). "Specific methanogenic activity test for anaerobic degradation of influents." *Applied Water Science*, 7(2): 535–542.

Kalogo, Y., and Verstraete, W. (2000). "Technical feasibility of the treatment of domestic wastewater by a CEPS-UASB system." *Environmental Technology*, 21(1): 55–65.

Ketheesan, B., and Stuckey, D. C. (2015). "Effects of hydraulic/organic shock/transient loads in anaerobic wastewater treatment: A review." *Critical Reviews in Environmental Science and Technology*, 45: 2693–2727.

Kida, Kenji, Kouhei Tanemura, and Yorikazu Sonoda. 1993. "Evaluation of the anaerobic treatment of sewage below 20°C by novel processes." *Journal of Fermentation and Bioengineering* 76 (6):510–514. doi: https://doi.org/10.1016/0922-338X(93)90250-C.

Kivistö, Anniina T., Alessandro Ciranna, Ville P Santala, and Matti T Karp. 2013. "Extremophilic microbes in biohydrogen production." *Current Biotechnology* 2 (4):345–359.

Lettinga, Gatze, Salih Rebac, Sofia Parshina, Alla Nozhevnikova, Jules B Van Lier, and Alfons JM Stams. 1999. "High-rate anaerobic treatment of wastewater at low temperatures." *Applied and Environmental Microbiology* 65 (4):1696–1702.

Lettinga, Gatze, Salih Rebac, and Grietje Zeeman. 2001. "Challenge of psychrophilic anaerobic wastewater treatment." *Trends in Biotechnology* 19 (9):363–370. doi: https://doi.org/10.1016/S0167-7799(01)01701-2.

Lettinga, G., S. Rebac, and G. Zeeman. (2001). "Challenge of psychrophilic anaerobic wastewater treatment." *Trends in Biotechnology*, 19(9): 363–370.

Lettinga, G., van Velsen, A. F. M., Hobma, S. W., de Zeeuw, W. J., & Klapwijk, A. (1980). "Use of the Upflow Sludge Blanket (USB) reactor concept for biological wastewater treatment, especially for anaerobic treatment." *Biotechnology and Bioengineering*, 22, 699–734. https://doi.org/10.1002/bit.260220402

Lew, Beni, Irina Lustig, Michael Beliavski, Sheldon Tarre, and Michal Green. 2011. "An integrated UASB-sludge digester system for raw domestic wastewater treatment in temperate climates." *Bioresource Technology* 102 (7):4921–4924. doi: https://doi.org/10.1016/j.biortech.2011.01.030.

Lew, Beni, Sheldon Tarre, M. Belavski, and Michal Green. 2004. "UASB reactor for domestic wastewater treatment at low temperatures: A comparison between a classical UASB and hybrid UASB-filter reactor." *Water Science and Technology* 49:295–301. doi: https://doi.org/10.2166/wst.2004.0865.

Lohani, SP, A Chhetri, and SN Khanal. 2015. "A simple anaerobic system for onsite treatment of domestic wastewater." *African Journal of Environmental Science and Technology* 9 (4):292–300.

Luostarinen, Sari, Wendy Sanders, Katarzyna Kujawa-Roeleveld, and Grietje Zeeman. 2007. "Effect of temperature on anaerobic treatment of black water in UASB-septic tank systems." *Bioresource Technology* 98 (5):980–986. doi: https://doi.org/10.1016/j.biortech.2006.04.018.

Maaz, Muhammad, Muhammad Yasin, Muhammad Aslam, Gopalakrishnan Kumar, AE Atabani, Mubbsher Idrees, Fatima Anjum, Farrukh Jamil, Rizwan Ahmad, and Asim Laeeq Khan. 2019. "Anaerobic membrane bioreactors for wastewater treatment: Novel configurations, fouling control and energy considerations." *Bioresource Technology* 283:358–372.

Madden, Pádhraig, Abdul M Al-Raei, Anne M Enright, Fabio A Chinalia, Dirk de Beer, Vincent O'Flaherty, and Gavin Collins. 2014. "Effect of sulfate on low-temperature anaerobic digestion." *Frontiers in Microbiology* 5:376.

Madsen, Michael, Jens Bo Holm-Nielsen, and Kim H Esbensen. 2011. "Monitoring of anaerobic digestion processes: A review perspective." *Renewable and Sustainable Energy Reviews* 15 (6):3141–3155.

Mahmoud, N., G. Zeeman, H. Gijzen, and G. Lettinga (2004). "Anaerobic sewage treatment in a one-stage UASB reactor and a combined UASB-Digester system." *Water Research* 38(9), 2348–2358.

McKeown, Rory M, Colm Scully, Therese Mahony, Gavin Collins, and Vincent O'Flaherty. 2009. "Long-term (1243 days), low-temperature (4–15 c), anaerobic biotreatment of acidified wastewaters: Bioprocess performance and physiological characteristics." *Water Research* 43 (6):1611–1620.

McKeown, Rory M., Dermot Hughes, Gavin Collins, Thérèse Mahony, and Vincent O'Flaherty. 2012. "Low-temperature anaerobic digestion for wastewater treatment." *Current Opinion in Biotechnology* 23 (3): 444–451. doi: https://doi.org/10.1016/j.copbio.2011.11.025.

Mrowiec, B.; Suschka, J. Anaerobic Wastewater Treatment Process. In Research and Application of New Technologies in Wastewater Treatment and Municipal Solid Waste Disposal in Ukraine, Sweden, and Poland; Plaza, E., Lewin, E., Eds.; KTH: Stockholm, Sweden, 2010; pp. 177–184.

Mukhtar, Salma, and Mehwish Aslam. 2020. "Biofuel synthesis by extremophilic microorganisms." In *Biofuels Production–Sustainability and advances in microbial bioresources*, 115–138. Springer.

O'Reilly, J, FA Chinalia, T Mahony, G Collins, J Wu, and V O'Flaherty. 2009. "Cultivation of low-temperature (15° C), anaerobic, wastewater treatment granules." *Letters in Applied Microbiology* 49 (4):421–426.

Petropoulos, Evangelos, Jan Dolfing, Russell J. Davenport, Emma J. Bowen, and Thomas P. Curtis. 2017. "Developing cold-adapted biomass for the anaerobic treatment of domestic wastewater at low temperatures (4, 8 and 15°C) with inocula from cold environments." *Water Research* 112:100–109. doi: https://doi.org/10.1016/j.watres.2016.12.009.

Sakcharoen, Thammananya, Chavalit Ratanatamskul, and Achara Chandrachai. 2021. "Factors affecting technology selection, techno-economic and environmental sustainability assessment of a novel zero-waste system for food waste and wastewater management." *Journal of Cleaner Production* 314:128103. doi: https://doi.org/10.1016/j.jclepro.2021.128103.

Saleh, Medhat MA, and Usama F Mahmood. 2004. "Anaerobic digestion technology for industrial wastewater treatment." Proceedings of the Eighth International Water Technology Conference, IWTC, Alexandria, Egypt.

Scully, Colm, Gavin Collins, and Vincent O'Flaherty. 2006. "Anaerobic biological treatment of phenol at 9.5–15 C in an expanded granular sludge bed (EGSB)-based bioreactor." *Water Research* 40 (20):3737–3744.

Seghezzo, L. 2004. Anaerobic treatment of domestic wastewater in subtropical regions. [internal PhD, WU, Wageningen University]. https://edepot.wur.nl/121555

Seghezzo, L., Castaneda, M. L., Trupiano, A. P., Gonzalez, S. M., Guerra, R. G., Torrea, A., Cuevas, C., Zeeman, G., and Lettinga, G. 2000. "Anaerobic treatment of pre-settled sewage in UASB reactors in subtropical regions (Salta, Argentina)." In: Proceedings of the VI Latin-American Workshop and Seminar on Anaerobic Digestion Volume I, Recife, Brasil, 2000 / Universitaria da UFPE. - Recife, Brasil : [s.n.], 2000.

Smith, Adam L., Steven J. Skerlos, and Lutgarde Raskin. 2013. "Psychrophilic anaerobic membrane bioreactor treatment of domestic wastewater." *Water Research* 47 (4):1655–1665. doi: https://doi.org/10.1016/j.watres.2012.12.028.

Smith, AL, SJ Skerlos, and L Raskin. 2015. "Anaerobic membrane bioreactor treatment of domestic wastewater at psychrophilic temperatures ranging from 15 C to 3 C." *Environmental Science: Water Research & Technology* 1 (1):56–64.

Smith AL, LB Stadler, L Cao, NG Love, L Raskin, and SJ Skerlos 2014. "Navigating wastewater energy recovery strategies: A life cycle comparison of anaerobic membrane bioreactor and conventional treatment systems with anaerobic digestion." *Environ Sci Technol* 48:5972–5981.

Tango, M. S. A., and M. R. Islam. 2002. "Potential of extremophiles for biotechnological and petroleum applications." *Energy Sources* 24 (6):543–559. doi: https://doi.org/10.1080/00908310290086554.

Tawfik, Ahmed, Manal Ali, Amal Danial, Shanshan Zhao, Fangang Meng, and Mahmoud Nasr. 2021. "2-biofuels (H2 and CH4) production from anaerobic digestion of biscuits wastewater: Experimental study and techno-economic analysis." *Journal of Water Process Engineering* 39:101736. doi: https://doi.org/10.1016/j.jwpe.2020.101736.

Tchobanoglus, George, Franklin Burton, and H David Stensel. 2003. "Wastewater engineering: Treatment and reuse." *American Water Works Association. Journal* 95 (5):201.

Tilche, A., and S. M. M. Vieira 1991. "Discussion Report on Reactor Design of Anaerobic Filters and Sludge Bed Reactors." *Water Science and Technology*, 24(8), 193–206. doi:10.2166/wst.1991.0225

Tsushima, Ikuo, Wilasinee Yoochatchaval, Hiroki Yoshida, Nobuo Araki, and Kazuaki Syutsubo. 2010. "Microbial community structure and population dynamics of granules developed in expanded granular sludge bed (EGSB) reactors for the anaerobic treatment of low-strength wastewater at low temperature." *Journal of Environmental Science and Health Part A* 45 (6):754–766.

Van Lier, Jules B, Nidal Mahmoud, and Grietje Zeeman. 2008. "Anaerobic wastewater treatment." *Biological wastewater treatment: Principles, modelling and design*, 415–456. London: IWA Pub. [edited by] Mogens Henze, Mark C.M. van Loosdrecht, George A. Ekama [and others].

Visvanathan, C., and Abeynayaka, A. 2012. "Developments and future potentials of anaerobic membrane bioreactors (AnMBRs)." *Membrane Water Treatment* 3, 1–23.

Xie Z., Z. Wang, Q. Wang, C. Zhu, Z. Wu 2014 "An anaerobic dynamic membrane bioreactor (AnDMBR) for landfill leachate treatment: Performance and microbial community identification. " *Bioresour. Technol.* 161: 29–39.10.1016/j.biortech.2014.03.014

Xu, Shengnan, Lei Zhang, Shengle Huang, Grietje Zeeman, Huub Rijnaarts, and Yang Liu. 2018. "Improving the energy efficiency of a pilot-scale UASB-digester for low temperature domestic wastewater treatment." *Biochemical Engineering Journal* 135:71–78. doi: https://doi.org/10.1016/j.bej.2018.04.003

5 Strategic Issues for Wastewater Reuse in Extreme Cold Conditions

Song Yan[a], Patrick Drogui[a], Rajeshwar D. Tyagi[b,c], and Jonathan W.C. Wong[b]
[a] Université du Québec, Québec, Canada
[b] Dongguan University of Technology, Dongguan, China
[c] BOSK Bioproducts, Québec, Canada

5.1 INTRODUCTION

In the past few decades, the increasing demand for water in drinking, irrigation, and industrial processes has created significant water stress issues. As a result, water scarcity and deteriorating water quality have become major challenges faced by many regions around the world (Jenssen, Vråle, and Lindholm 2007; Islam, Reinstädtler, and Eslamian 2015; Diaz-Elsayed et al. 2019; Van Rossum 2020; Nagpal et al. 2021; Knutsson and Knutsson 2021; Samuel et al. 2022). Wastewater recycling and reuse is a rational and practical approach that contributes to environmental protection worldwide. By reusing water, several benefits can be achieved, including increasing water availability, reducing the demand for freshwater resources in agricultural irrigation and industrial processes, promoting sustainable management of water resources in different climatic conditions, and supporting environmental and public health protection (Shelef and Azov 1996, Vazquez-Montiel, Horan, and Mara 1996; El Moussaoui et al. 2019; Van Rossum 2020; Wang et al. 2020; Knutsson and Knutsson 2021).

Climatic conditions are a crucial factor in regional water use, reuse, and management issues. In cold climate regions, these issues become even more challenging, as people face significant obstacles in maintaining natural surface and groundwater systems essential for sustainable water management (Islam, Reinstädtler, and Eslamian 2015).

This chapter presents a comprehensive analysis of the various applications of wastewater reuse, including in agriculture under cold climate conditions, the health risks associated with agricultural wastewater reuse, the use of wastewater as fertilizers, as well as its reuse as toilet flushing water and as a source of energy. Additionally, the factors driving the adoption of wastewater reuse and the necessary considerations for ensuring sustainable management are also discussed.

5.2 WASTEWATER REUSE IN AGRICULTURE

5.2.1 Guidelines for the Safe Reuse of Wastewater in Agriculture

The World Health Organization (WHO) published three guidelines in 1973, 1989, and 2006 for the safe use of wastewater in agriculture (Carr 2005; Mara and Kramer 2008). These guidelines were intended to support the establishment of national standards and regulations. The latest WHO guidelines published in 2006 (WHO 2006a, 2006b) are essentially a code of good management practices to ensure that when wastewater is used in agriculture (mainly for irrigating crops, including food crops that are or may be eaten uncooked), it is used safely and with minimal risks to health. To

DOI: 10.1201/9781003376415-6

reduce the health risks resulting from human exposure to pathogens in the wastewater, the guidelines focus on health-based targets, instead of water quality standards, and offer various combinations of risk management options for meeting them. The Food and Agriculture Organization of the United Nations (FAO) (FAO 2017) also has two guidelines for the application of water reuse in irrigation. The US Environmental Protection Agency (EPA) also released four wastewater reuse guidelines in 1980, 1992, 2004, and 2012. These guidelines are considered as the bases for generating the regulations and guidelines in different countries worldwide. The government departments normally involved in this decision-making process are ministries or departments of health, water, environment and finance, including the part of government responsible for food safety.

For example, in The 2006 WHO Guidelines for Wastewater and Greywater Use in Agriculture: A Practical Interpretation (Mara and Kramer 2008): Reuse of Greywater, unrestricted irrigation with greywater is beneficial for increasing crop yields, and the pathogen levels are relatively low (Jackson et al., 2006; WHO 2006b). Compared to domestic wastewater (i.e., greywater and blackwater combined), greywater poses lower health risks, as the pathogen numbers are significantly lower due to the smaller fecal load present (a cross-contamination load of ~0.04 g feces per person per day enters greywater, resulting in a greywater quality of ~10^4–10^5 *E. coli* per 100 mL, as opposed to 10^7–10^8 per 100 mL for domestic wastewater; WHO 2006b). According to quantitative microbial risk analyses (QMRA) studies reported in WHO (2006b), a 1.6- to 2.9-log unit reduction is required for protozoan pathogens (Cryptosporidium, Giardia) and a 2.3- to 3.3-log unit reduction for viral pathogens (rotavirus) to prevent exceeding the tolerable additional disease burden of 10^{-6} Disability-Adjusted Life Years (DALY) loss per person per year (pppy). These reductions can be achieved, as in the case of wastewater, through a combination of treatment and post-treatment health protection control measures. However, it's worth noting that little to no treatment is required as these pathogen reductions are achievable through natural pathogen die-off and produce washing/disinfection/peeling. Nonetheless, retaining greywater in a tank for a few hours can be beneficial to remove scum and readily settleable solids.

5.2.2 Benefits of Agricultural Reuse

Treated wastewater used in agriculture irrigation could be a realistic solution for the shortage of fresh water. Jenssen (2004) conducted a study of a combined vertical flow biofilter followed by a horizontal flow wetland filter for advanced greywater treatment. The results showed that the total area requirement is 1 m^2/person and the effluent meets European swimming water standards with respect to indicator bacteria, and WHO drinking water standards with respect to nitrogen. The low area requirement of the system and the high effluent quality facilitates use in urban settings, and discharge to small streams, open waterways, or irrigation or groundwater recharge (Jenssen 2004).

Operating constructed wetlands (CWs) in cold temperatures can be a challenge, as it requires significant investment and maintenance costs (Gao and Hu 2012; Wu et al. 2014). Gao and Hu (2012) found that using a greenhouse-structured wetland can increase the temperature in colder climates, leading to improved contaminant removal efficiency. To treat sewage for reuse in irrigation in rural China, a full-scale wastewater treatment plant was developed using a novel greenhouse-structured, horizontal subsurface flow constructed wetland (HFCW) with bio-contact oxidation (BCO) pre-treatment. Results showed that the average annual concentrations of chemical oxygen demand (COD), ammonia nitrogen (NH_3-N), and total phosphorus (TP) in the final effluent were 22.40 (±6.84) mg/L, 8.58 (±1.48) mg/L, and 2.09 (±0.28) mg/L, respectively. The greenhouse insulation helped overcome the limitations of using only subsurface wetlands and improved pollutant removal efficiency in winter. Probability analysis suggests that the BCO–HFCW combined system may be a viable alternative for wastewater treatment and reuse in cold regions with temperatures below 0°C (Gao and Hu 2012). This study provides a good example of successful wastewater treatment and reuse in cold climate conditions.

Wu et al. (2011) demonstrated that an integrated household constructed wetland (CW) system had significant potential to effectively remove chemicals in cold climates, with no reduction in treatment efficiency observed during the winter, especially in northern China. Zhang et al. (2015) developed a natural heating system that utilized composting pig manure to raise the temperature of CWs in cold climates (Zhang et al. 2015).

5.2.3 Health Risks of Agricultural Wastewater Reuse

Wastewater disinfection is important to protect human and ecosystem health. Chemical disinfection is commonly used as the final stage in wastewater treatment to reduce the release of pathogenic microorganisms into the environment. Chlorine, either in the form of chlorine gas (Cl_2) or sodium hypochlorite solution (NaOCl), is the most widely used disinfection method due to its high effectiveness, ease of use, and relatively low capital and operational costs (Hassaballah et al. 2020; US Environmental Protection Agency 1999).

Peracetic acid (PAA) could be used as a disinfectant for wastewater and has the advantage of being less toxic to aquatic species and forming fewer known disinfection byproducts than traditionally used chlorine (Hassaballah et al. 2020).

The disinfection efficiency of PAA and chlorine was compared (Hassaballah et al. 2020) for four microorganisms: two fecal indicator bacteria (*E. coli* and *Enterococcus* spp.), a virus (somatic coliphage), and a protozoan (*Cryptosporidium parvum*). Experiments were conducted in two temperatures (room temperature and 4°C). At a winter-relevant temperature (4°C), NaOCl was more effective against *E. coli* and *Enterococcus* spp. than PAA regardless of exposure, while both PAA and NaOCl achieved comparable somatic coliphage inactivation. The results revealed the use of PAA as an alternative wastewater disinfectant to chlorine and as a complementary disinfectant to UV (Hassaballah et al. 2020)

A four-year field trial (2016–2019) was carried out near a wastewater treatment plant in eastern Finland (Outokumpu) to investigate the effect of the processed wastewater (WW) on biomass production as well as the nutrients uptake capability (mainly N and P) by a willow variety (*Salix schwerinii*). Results indicated that WW irrigation expressively increased the willow diameter growth and biomass yield around 256% and 6510%, respectively, compared to the control treatment site (without WW). The willow was also able to accumulate approximately 41–60% of the N and 32–50% of the P in two years (2018–2019). Overall, willow showed a total 20% mortality rate under WW irrigation throughout the growing periods (2017–2019) as compared to control (39%). The results demonstrated that willow had the potential to control eutrophication (reducing nutrients load) from the wastewater with the best survival rate and could provide high biomass production for bioenergy generations in cold climatic conditions (Mohsin et al. 2021).

When considering the use of effluent for irrigation, it is essential to not only meet the required physicochemical quality standards but also to consider microbial parameters. Pathogens present in the effluent can pose a significant threat to public health. To safeguard public health, it is recommended to include post-treatment and advanced treatments, such as UV disinfection (Vazquez-Montiel, Horan, and Mara 1996; Scott et al. 2009; Baghapour, Nasseri, and Djahed 2013; Ghasemi-Zaniani et al. 2017; Hassaballah et al. 2020).

A recent study (Heaven et al. 2007) examined the use of waste stabilization ponds in extreme continental climates like those found in central Asia, where low precipitation and high summer evaporation rates pose challenges for water reuse. Using a simple model, researchers predicted water availability, biochemical oxygen demand (BOD), and fecal coliform removal rates for different configurations and operating regimes. The results showed that significant amounts of water could be saved for irrigation or replenishing rivers and aquifers, provided standard designs are modified for these climates. Three case studies of evaporation pond systems in Kazakhstan further revealed their potential for conversion into full biological treatment systems, allowing for greater water conservation and reuse flexibility (Heaven et al. 2007).

5.2.4 Wastewater Reuse as Fertilizers

Treated wastewater has organic matter (OM) and nutrients such as N, P, K, and Fe, Mn, Zn, and Cu, and a significant amount of organic matter (OM) that are useful to the plants; therefore, it can be used as fertilizer. Wastewater reuse has been proven to improve crop yield and result in the reduced amount and cost of fertilizers in agriculture (Tierney, Sullivan, and Larkin 1977; Sengupta, Nawaz, and Beaudry 2015; Zhang, Jiang, and Deng 2017; Antille et al. 2017).

Completely decentralized wastewater treatment systems in urban areas where the blackwater fractions (urine and fecal matter) is possible to be reused for fertilizer and potentially energy production (Jenssen et al. 2003; Jenssen 2004). The water from kitchen sinks and showers (greywater) is treated locally in compact low maintenance systems that constitute attractive landscape elements (Jenssen and Vråle 2003). These systems can coexist with decentralized water supply. Greywater may contain more than 50% of the organic matter and, surprisingly, a relatively high content of bacteria and even virus (Ottoson 2003). If greywater is discharged to smaller streams in urban areas or used for irrigation or groundwater recharge, advanced treatment that reduce organic matter and the hygienic parameters is needed. At Klosterenga, in the capital of Norway, Oslo, the greywater is treated in an advanced nature-based greywater treatment system in the courtyard of the building. The system consists of a septic tank, pumping to a vertical down-flow single-pass aerobic biofilter followed by a subsurface horizontal-flow porous media filter (Jenssen 2004).

A feasibility study of sustainable urban wastewater reuse in the agroforestry domain of Marrakesh, Morocco, was performed (El Moussaoui et al. 2019). The zone of Marrakesh is located in the arid bioclimatic stage characterized by scarce precipitation, grouped during the cold season, from September to May. January was the coldest month, with a monthly average temperature of 10.96°C (El Moussaoui et al. 2019). Treated urban wastewater reuse is a sustainable and promising strategy to face water scarcity, enhance soil fertility, preserve natural resources, develop local products, and improve the living conditions of agriculture and farmers (El Moussaoui et al. 2019).

Land disposal of sewage sludge and effluent is becoming a common practice in the United States. The fertilizer content and humus value of such wastes are useful for agricultural purposes, and the recycling of sewage onto the land eliminates many of the stream pollution problems. The potential exists for crops grown in such irrigated soil to be contaminated by viruses that may be present in the sewage. The longest period of survival was during the winter, when virus was detected after 96 days (Tierney, Sullivan, and Larkin 1977).

Anaerobic treatment in small, decentralized units has been considered uneconomical in Norway. This is partly due to safety regulations, but also the cold climate that demands better insulation and more sophisticated systems than in warm climates. Nevertheless, anaerobic processes are attractive due to the energy quality of gas being superior to heat, and less energy is needed to operate an anaerobic process. Therefore, researchers have started considering the use of small-scale anaerobic reactors in cold climate conditions (Jenssen, Vråle, and Lindholm 2007).

Use of nitrogen- and phosphorus-based synthetic fertilizers shows an increasing trend, but this has led to a large-scale influx of reactive nitrogen in the environment, with serious implications on human health and the environment. On the other hand, phosphorus, a nonrenewable resource, faces a serious risk of depletion. Therefore, recovery and reuse of nitrogen and phosphorus is highly desirable. For nitrogen recovery, an ion exchange/adsorption-based process provides concentrated streams of reactive nitrogen. Bioelectrochemical systems efficiently and effectively recover nitrogen as NH_3 (g) or $(NH_4)_2SO_4$. Air stripping of ammonia from anaerobic digestate has been reported to recover 70–92% of nitrogen. Chemical precipitation can remove 80–99% P in wastewater streams and recover it in the form of fertilizer (struvite). Acid hydrolysis can convert recovered P into usable phosphoric acid and phosphate fertilizers. Physical-chemical adsorption and ion exchange media can reduce P to trace or nondetectable concentrations, with minimal waste production and high reusability. There are good methods on recovery of nitrogen and phosphorus from wastewater (Sengupta, Nawaz, and Beaudry 2015).

Recovery and reuse of P from wastewater treatment systems as agricultural fertilizer presents an important and viable target for P waste reduction and recycling.

Field-scale experiments (Antille et al. 2017) in four crop seasons established the agronomic performance of biosolids-derived organomineral fertilizers (OMF) for winter wheat (*Triticum aestivum* L.) production in England. Two OMF formulations (OMF_{10} 10:4:4 and OMF_{15} 15:4:4) were compared with urea and biosolids granules (≈5:6:0.2) to determine crop responses and fertilizer effects on soil chemical properties. Average grain yields with OMF_{10} and OMF_{15} were higher than with biosolids granules, but lower than with urea ($P < 0.05$). It revealed that agronomic efficiency of organomineral fertilizers was higher than biosolids granules; organomineral fertilizers reduced the risk of soil P buildup compared with biosolids. Heavy metals in soil were unaffected by fertilizer treatment and lower than permissible limit values. The use of OMF for winter wheat production was revealed to be a sustainable approach to recycling biosolids to land (Antille et al. 2017). In cold climate conditions, wastewater is reused as fertilizer as well.

5.3 WASTEWATER REUSE AS TOILET FLUSHING WATER (IN COLD CONDITIONS)

The number of schemes for recycling water for toilet flushing and other non-potable urban uses is growing. Nowadays, one of the most interesting issues for wastewater recycling is the on-site treatment and reuse of greywater.

The efficiency of a compact submerged membrane bioreactor (SMBR) system to treat real greywater in a single house in Crete, Greece, was examined (Fountoulakis et al. 2016). Greywater was collected from a bathtub, shower, and washing machine containing significant amounts of organic matter and pathogens. COD removal was approximately 87%. TSS were reduced from 95 mg L^{-1} in the influent to 8 mg L^{-1} in the effluent. The removal of anionic surfactants was about 80%. Fecal and total coliforms decreased significantly, due to rejection, by the membrane, used in the study. Overall, the SMBR treatment produces average effluent values that would satisfy international guidelines for indoor reuse applications such as toilet flushing (Fountoulakis et al. 2016). This is available in cold climate conditions.

Reuse of sewage effluent has the potential to both minimize the disposal of water to the environment and reduce the demand on fresh water supplies. In Canberra the Australian national capital city, trials were undertaken to demonstrate the practicalities of reuse. The first, referred to as "water mining," consists of a small treatment plant constructed within the urban environment that extracts flows from an adjacent sewer and uses the treated effluent for the irrigation of nearby playing fields. In the second project, six houses have been disconnected from the city sewerage system and each provided with its own treatment plant. Effluent is stored in tanks on-site and used for garden and lawn irrigation and toilet flushing. The third project also involves a house disconnected from the sewerage system. The blackwater and greywater strains were kept separate. A low-cost process has been devised for the treatment of the greywater stream with effluent, again used for irrigation and toilet flushing (Neal 1996).

Knutsson and Knutsson (2021) investigated a greywater reuse scheme and the impact of system design and configuration on water and energy savings. A simulation model was created based on real, disaggregated water consumption data that predicts the reuse potential. Three scenarios were investigated: (1) greywater collection from the bathroom and reuse for toilet flushing; (2) greywater collection from bathroom sinks and showers, and reuse as hot water for sinks and showers; and (3) a combination of (1) and (2) where greywater collection from bathroom sinks and showers is used for toilet flushing, sinks, and showers. The results showed hot water reductions between 55.6 and 58.2%, while cold water reductions ranged from 5.8 to 30.6%. Reductions in energy for producing hot water between 43.5 and 46.8% were observed (Knutsson and Knutsson 2021).

Greywater is a potential water source for reducing water demand. Accordingly, a review was undertaken of greywater reuse regulations and guidelines within the 50 United States (Yu et al. 2013).

Major issues considered included acceptability for greywater segregation as a separate wastewater stream, allowable greywater storage, on-site treatment requirements, and permitted greywater use applications. Existing regulations and plumbing codes in the different states suggest that there are impediments to overcome but also potential incentives for greywater reuse. It is encouraging that regulations in 29 states promote safe greywater reuse, but there are also inconsistencies between plumbing codes and other regulations within and among the 50 states. Impediments to greywater reuse include disallowances of greywater segregation or collection, and restriction of greywater reuse to mostly subsurface irrigation with limited indoor use permission. Ease on restrictions and guidelines to promote the development of low-cost and proven treatment technologies is needed to promote greywater reuse (Yu et al. 2013).

Greywater treatment and reuse for non-potable purposes in urban areas has become a widely researched topic to reduce the burden on freshwater resources. Lakho et al. (2021) reported the use of a green wall, called Total Value Wall (TVW), for treating greywater and reusing the effluent for toilet flushing. Initially, the effectiveness of (mixtures of) different substrates, i.e. lava, lightweight expanded clay aggregates, organic soil, and biochar was investigated by means of column tests. All substrates were first examined for hydraulic characteristics and later on the columns were fed with synthetic grey wastewater and followed up in terms of removal efficiency of COD and detergents. The mixture consisting of lava (50%), organic soil (25%), and biochar (25%) proved to be optimal both in terms of percolation rates and removal efficiencies, and was thus selected for the full-scale system. The full-scale TVW of 14.4 m^2 was installed at a terraced house in Ghent (Belgium) and was loaded with greywater at 100 L per day. Influent and effluent quality were routinely monitored by grab sampling, water savings were monitored by means of flow meters, and electricity consumption was also accounted for. The TVW was further equipped with sensors that measure temperature, particulate matter (PM_{10}), and CO_2 in the air. The full-scale system obtained effluent concentrations of 13 $mg{\cdot}L^{-1}$ TSS, 91 $mg{\cdot}L^{-1}$ COD, and 5 $mg{\cdot}L^{-1}$ BOD_5. Ammonium and total coliforms were removed with removal rates of 97% and 99% (2 log units), respectively. Available data from the temperature sensors have clearly demonstrated the additional benefit of the TVW as an insulating layer, keeping the heat outside on warmer days, and keeping the heat inside on colder days. Overall, this study demonstrated that the TVW is a sustainable system for greywater treatment and reuse (Lakho et al. 2021).

Examples of recycling water for toilet flushing from around the world include well-established, municipal scale (in the United States and Japan) and newer, local community scale (in Europe and Australia) (Lazarova, Hills, and Birks 2003).

Lazarova, Hills, and Birks (2003) summarized the current non-potable, urban use of reclaimed water with particular reference to toilet flushing. They compared water quality standards for reclaimed water, the volumes of water required for toilet flushing, and the qualities of greywater and domestic sewage that had previously been used for reuse. It was demonstrated that the interest in water reuse is growing steadily, not only in acknowledged water deficient areas but also in countries that have not historically appeared to have a water supply problem (Lazarova, Hills, and Birks 2003).

The "Toronto Healthy House" was a demonstration project initiated by a design competition by Canada Mortgage and Housing Corporation (CMHC) in 1991 (Leung 2004). In the report by Ada Leung (2004), the house water supply and reuse water system included a cistern beneath the backyard deck connected to a filtration system that collects rain and greywater for recycling and reuse within the house. Approximately 80% of the water used in the home is through recycling.

Reuse in London, Ontario, includes a 3M office and Fanshawe College. The 3M office uses cistern water from the basement sump weeping tile system for flushing toilets. This effort reduced building potable water consumption by 25% annually for the 600 employees at the site (about 200 m^3 per month since 2017).

In British Columbia, examples of buildings with internal wastewater reuse and non-potable reuse systems are the Vancouver Convention Centre (VCC) West building (Vancouver, BC); and the

University of British Columbia Centre for Interactive Research in Sustainability (CIRS) building (Vancouver, BC) (Vassos 2014). The CIRS building collects rainwater from the roofs of the building and stores it in a cistern. Sprinkler irrigation can use the backup cistern water as a secondary supply of water. The rainwater is filtered and disinfected on-site and distributed through the building for potable water applications. Wastewater from the building is treated on-site and used for toilet flushing and irrigation within the building. The Vancouver Convention Centre (VCC) West building has a blackwater treatment plant that reuses greywater and blackwater for washrooms toilet flushing and rooftop irrigation during warmer weather. Blackwater generally refers to wastewater originating from sanitary sources (e.g. toilets, urinals, and bidets), as well as drainage from food preparation and utensil cleaning activities (e.g. kitchen sinks and dishwashers) (Vassos 2014).

5.4 WASTEWATER REUSE AS A SOURCE OF ENERGY (IN COLD CONDITIONS)

Wastewater source heat pumps (WWSHPs) have become increasingly popular in many countries due to their advantages of relatively higher energy utilization efficiency and environmental protection (Shen et al. 2018).

The EU Directive 2018/2001 recognized wastewater as a renewable heat source. Wastewater from domestic, industrial, and commercial developments maintains considerable amounts of thermal energy after discharging into the sewer system. It is possible to recover this heat by using technologies like heat exchangers and heat pumps; and to reuse it to satisfy heating demands (Nagpal et al. 2021). It is generally accepted that a global sustainable energy system will be a combination of many different renewable resources. Hence, a movement toward maximizing utilization of biologically derived energy coupled with other renewable energy sources has recently become of more interest to wastewater facility management as part of the global renewable energy portfolio. Although the challenges faced by engineers are imposing, the industry has responded strongly in recent decades with novel and valuable research efforts and innovative engineering approaches to applying these concepts to advanced performance in the field. These technological advances can be utilized for generation of renewable energy (Whitlock et al. 2009).

In recent times, the practice of using a horizontal ground heat exchanger to extract heat from sewage pipes has emerged as a promising renewable energy option, aimed at reducing reliance on fossil fuels. A recent study (Dacquay et al. 2020) proposed an innovative design for a ground heat exchanger, which harnesses heat from the soil and sewage in a pipe while it's transported to a wastewater treatment plant. The researchers used a model with constant inlet fluid temperatures to assess heat transfer through convective pipe flow and conductive phase change in the soil. The results indicated a maximum temperature change of only 0.10°C in the surrounding soil over 25 years during the heating season in Winnipeg, Manitoba. The key factors affecting the system's performance were the depth of the system, sewage level, and the thermal properties of the high-density polyethylene pipe. The system's sustainability was not compromised by its depth, as thermal balancing occurred between climatic, subsurface, and sewage heat fluxes. The system performed sustainably at 50% and 75% sewage pipe capacity. The impact of the high-density polyethylene pipe's thermal properties on thermal performance was found to be insignificant (Dacquay et al. 2020).

Constructed wetlands (CW) are an eco-technology that is widely used for wastewater reclamation. They are ideal for achieving water security, energy harvesting, and environmental services. Recent trends have shown additional research interventions for better expansion of the technology, such as energy harvesting to make the system a net energy producer by coupling CW with microbial fuel cells (CW-MFC), and improved operation under challenging climatic conditions. However, low-temperature challenges of the technology need to be addressed, and strategies to achieve satisfactory low-temperature performance should be assessed. It would also be beneficial to conduct a comparative evaluation of the technology with other treatment processes using life cycle assessment (LCA) with a cradle-to-grave approach, considering alternative substrates, energy harvesting, macrophyte use, and disposal options, which could further boost the technology penetration (Kataki et al. 2021).

Two advanced municipal wastewater treatment plants in Austria, equipped with nutrient removal technology, have achieved energy self-sufficiency (Nowak, Enderle, and Varbanov 2015). These plants have generated up to 180% more energy than they require. By switching to cooling water regeneration as the heat source for heat pumps in district heating, electricity savings of up to 45% can be realized. However, it's important to avoid negative environmental impacts, such as inadequate wastewater treatment or the release of methane gas into the atmosphere (Nowak, Enderle, and Varbanov 2015).

5.5 OTHER REUSE (INDUSTRIAL APPLICATIONS)

The city of Edmonton, located in Alberta, Canada, partnered with Suncor on a wastewater reuse project and pipeline to support a $2 billion expansion and upgrade to the Suncor refinery. The Gold Bar Wastewater Treatment Plant treats wastewater from about 700,000 people and has a daily capacity of treating 310 million liters of wastewater. The city of Edmonton, sole shareholder of the Utility Company (EPCOR), is operating the membrane facility. Suncor constructed a 5.5-km, 460-mm water pipeline between the refinery and EPCOR's Gold Bar Wastewater Treatment Plant. The partnership between Suncor and EPCOR ensured a consistent, high-quality water supply for the refinery, providing 15 million liters of water each day. The refinery reused the water for cooling towers and hydrogen plants, and also for the boiler water, which went through an ultrafiltration and reverse osmosis system (Van Rossum 2020).

Water reuse is a significant alternative water resource to alleviate the increasing demands for water resources and to address water quality deterioration issues. Lessons learned from water reuse practices in Canada can be beneficial to other countries and regions, particularly those in arid areas that suffer from severe water environment problems or remote locations (Van Rossum 2020).

5.6 CASE STUDIES

Unconventional water reclamation, recycling, and reuse have been identified as the major approaches for addressing water scarcity in China.

Wang et al. (2019) evaluated the technical and economic feasibility of water reclamation plants (WRP) based on an anaerobic-anoxic-oxic membrane bioreactor (A2O-MBR) system for unconventional water resource (domestic wastewater and rainwater) treatment and reuse in a green building residential community. During a year-long operation and evaluation process, average COD, BOD_5, total nitrogen (TN), and total phosphorus (TP) removal efficiencies averaged $83.7 \pm 3.2\%$, $91.6 \pm 2.6\%$, $69.0 \pm 6.5\%$, and $74.5 \pm 5.5\%$, respectively. The effluent quality of wastewater treated by the A2O-MBR system complied with the water quality standards for reuse for the green building residential community or for direct discharge. Typical seasonal fluctuations of temperature, COD loading rate, BOD_5/COD, and COD/TN were the key factors affecting pollutants removal efficiencies, and the membrane fouling rate was accelerated at low temperatures. The total cost of capital and operation was 0.406 $\$/m^3$; a cost-benefit analysis incorporating both capital and operating expenditures showed that more than 60% of the reclaimed water is reused in the green building, and the full-scale WRP exhibits a positive net present value. These results demonstrated that A2O-MBR is an efficient and profitable technology for nontraditional water resource recycling in green building residential communities in terms of technical and economic feasibility (Wang et al. 2019).

The recovery and reuse of energy from sewage and treated water is a desirable plan for a sewerage system in an urban area due to its large heat capacity. Funamizu et al. (2001) presented the characteristics of heat energy in wastewater reuse plans and some experiences in Japan. Full-scale reuse projects were investigated for heating and cooling in the Tokyo Metropolitan Districts and a project for melting snow in Sapporo City. The experience of Tokyo showed that setting the heat pumps near the demand points and technical developments of equipment were key factors in the success of the project. The equipment needed to prevent system clogging, corrosion, and decrease in the heat transfer efficiency. In the project for melting snow in Sapporo, the key factor in public acceptance was the multipurpose use of the sewerage system for both melting snow in winter and retaining rainwater in summer (Funamizu et al. 2001).

Worldwide, high energy consumption affects the wastewater industry and it is second in cost only to manpower for most wastewater treatment plants (WWTPs). Energy prices are on the rise; the operating costs of WWTPs is increasing too. Energy production, usage, and saving in WWTP have been discussed while one WWTP in Ireland was examined in order to (1) identify energy/resource recovery strategies, (2) explore the potential applications in the plant, (3) promote the commercial and large-scale application at WWTPs (Awe, Liu, and Zhao 2016).

In order to improve the comprehensive energy utilization efficiency of energy systems with sewage resources, Wang et al. (2020) proposed an integrated energy system of wastewater reuse and an optimal operation model. The energy system operation control strategy was based on sewage source heat pump and biogas trigeneration energy production. Simulation results showed that compared with no optimization, the operating cost of the system could be reduced more than 9%, and the renewable energy utilization rate could be increased more than 5%. After the thermoelectric ratio adjustment, the results of system operation optimization could be further improved (Wang et al. 2020).

5.7 CHALLENGES IN COLD CONDITIONS

Wastewater is a readily available but often underutilized resource that could be used for irrigation and other purposes while conserving freshwater for future generations. Increasingly, wastewater management is recognized as an essential and cost-effective strategy in areas with water scarcity. However, the local climatic conditions can pose a significant challenge for water resources and wastewater reuse. Therefore, it is crucial to investigate the impact of cold climate conditions on the reuse of wastewater for agricultural and other purposes.

To develop sustainable wastewater treatment and reuse systems, it is essential to consider them in a broader context. Evaluating the sustainability of such systems and selecting appropriate ones for specific conditions requires taking into account recycling, reuse, and social issues. Achieving this requires a multidisciplinary approach involving engineers, social scientists, economists, biologists, health officials, and the public. The nutrients in domestic wastewater and organic household waste are almost sufficient to produce enough food for the global population, and recycling and energy aspects are crucial for sustainable system design. Water-saving and urine-diverting toilets can significantly reduce water consumption. Separate collection of toilet waste (blackwater) and urine allows for co-treatment with organic household waste to produce energy and hygienic fertilizer. Greywater from showers, sinks, and kitchens can also be treated in various systems for irrigation, groundwater recharge, or potable water production. By utilizing such systems, more than 90% water savings are possible, and source separation systems can produce near-zero emissions while opening up exciting urban applications (Jenssen, Vråle, and Lindholm 2007; Islam, Reinstädtler, and Eslamian 2015; Uddin et al. 2016; Wang et al. 2017; Kinsley, Kennedy, and Crolla 2017; Khurelbaatar et al. 2021).

The perception of users is a crucial factor when it comes to the use of recycled wastewater. Studies on recycled water schemes have generally found a positive attitude toward its use for toilet flushing. However, its acceptability decreases for more "personal" uses such as watering vegetables. Users expect operators of such systems to maintain appropriate quality and operate the systems safely. As water resources become more limited and debates around potable reuse continue, the use of recycled water for specific non-potable urban purposes such as toilet flushing will become increasingly attractive (Lazarova, Hills, and Birks 2003).

5.8 CONCLUSIONS AND PERSPECTIVES

Wastewater reuse is becoming an increasingly important strategy for conserving water resources and mitigating water scarcity in many regions of the world. However, in extreme cold conditions, such as those found in northern latitudes, wastewater reuse faces unique challenges that must be addressed to ensure its feasibility and effectiveness. This chapter has reviewed some of the key strategic issues for wastewater reuse in extreme cold conditions, including the potential impacts of salt contamination, the need for decentralized treatment systems, and the importance of stakeholder engagement and policy support.

The impact of salt contamination is a critical concern for wastewater reuse in extreme cold conditions, as deicing salts are commonly used to maintain road safety in these regions. These salts can enter the wastewater system through runoff and may cause corrosion of infrastructure and harm to aquatic life, if not properly treated. To mitigate these impacts, careful selection and design of treatment technologies is necessary to remove salt and other contaminants from wastewater, such as membrane filtration, ion exchange, and biological treatments. Additionally, regulatory frameworks and guidelines should be established to ensure that wastewater reuse is safe and environmentally sustainable.

Decentralized treatment systems are another key consideration for wastewater reuse in extreme cold conditions. These systems are necessary to address the logistical challenges of transporting and treating wastewater in remote areas where centralized treatment plants may be impractical or cost prohibitive. Decentralized systems, such as package plants or constructed wetlands, can be designed to meet the specific needs of a community or region, and can be more cost-effective and environmentally sustainable than traditional centralized systems. However, challenges such as limited space and harsh weather conditions must be considered in the design and implementation of decentralized treatment systems.

Stakeholder engagement and policy support are also crucial components of successful wastewater reuse in extreme cold conditions. Community members, local governments, and other stakeholders must be involved in the planning and implementation of wastewater reuse projects to ensure that their needs and concerns are addressed. Additionally, policies and regulations that support wastewater reuse, such as incentives for water conservation and reuse, can help to promote the adoption of wastewater reuse technologies and practices.

Looking ahead, several future perspectives should be considered for wastewater reuse in extreme cold conditions. First, the development and implementation of innovative and cost-effective treatment technologies should be prioritized, particularly those that can address salt contamination and other unique challenges posed by extreme cold conditions. Additionally, research should be conducted to better understand the environmental impacts of wastewater reuse in extreme cold conditions, as well as the potential benefits and risks of different treatment technologies. Finally, stakeholder engagement and policy support should continue to be a focus for promoting the adoption of wastewater reuse in extreme cold conditions.

REFERENCES

Antille, Diogenes L., Richard J. Godwin, Ruben Sakrabani, Saman Seneweera, Sean F. Tyrrel, and A. Edward Johnston. 2017. "Field-Scale evaluation of biosolids-derived organomineral fertilizers applied to winter wheat in England." *Agronomy Journal* 109 (2):654–674.

Awe, Olumide Wesley, Ranbin Liu, and Yaqian Zhao. 2016. "Analysis of energy consumption and saving in wastewater treatment plant: Case study from Ireland." *Journal of Water Sustainability* 6 (2):63–76.

Baghapour, Mohammad A., Simin Nasseri, and Babak Djahed. 2013. "Evaluation of Shiraz wastewater treatment plant effluent quality for agricultural irrigation by Canadian water quality index (CWQI)." *Iranian Journal of Environmental Health Science & Engineering* 10 (1):27. doi: https://doi.org/10.1186/1735-2746-10-27.

Carr, R. (2005). "WHO guidelines for safe wastewater use-more than just numbers." Irrigation and Drainage, 54, 103–111.

Dacquay, Connor, Hartmut M. Holländer, Miroslava Kavgic, Pooneh Maghoul, Hongwei Liu, and Hikari Fujii. 2020. "Evaluation of an integrated sewage pipe with ground heat exchanger for long-term efficiency estimation." *Geothermics* 86:101796. https://doi.org/10.1016/j.geothermics.2019.101796.

Diaz-Elsayed, Nancy, Nader Rezaei, Tianjiao Guo, Shima Mohebbi, and Qiong Zhang. 2019. "Wastewater-based resource recovery technologies across scale: A review." *Resources, Conservation and Recycling* 145:94–112. https://doi.org/10.1016/j.resconrec.2018.12.035.

El Moussaoui, Tawfik, Said Wahbi, Laila Mandi, Salvatore Masi, and Naaila Ouazzani. 2019. "Reuse study of sustainable wastewater in agroforestry domain of Marrakesh city." *Journal of the Saudi Society of Agricultural Sciences* 18 (3):288–293.

FAO. 2017. AquaStat, Food and Agriculture Organization of the United Nations. Available online: http://www.fao.org/nr/water/aquastat/data/query/results.html.

Fountoulakis, M.S., N. Markakis, I. Petousi, and T. Manios. 2016. "Single house on-site grey water treatment using a submerged membrane bioreactor for toilet flushing." *Science of the Total Environment* 551:706–711.

Funamizu, N., M. Iida, Y. Sakakura, and T. Takakuwa. 2001. "Reuse of heat energy in wastewater: Implementation examples in Japan." *Water Science and Technology* 43 (10):277–285.

Gao, Da-Wen, and Qi Hu. 2012. "Bio-contact oxidation and greenhouse-structured wetland system for rural sewage recycling in cold regions: A full-scale study." *Ecological Engineering* 49:249–253. https://doi.org/10.1016/j.ecoleng.2012.08.012.

Ghasemi-Zaniani, Mahboobe, Saeid Eslamian, Kaveh Ostad-Ali Askari, and V. Singh. 2017. "Irrigation with waste water treated by constructed wetlands." *International Journal of Research Studies in Agricultural Sciences* 3:18–34.

Hassaballah, Abdulrahman H., Tanmay Bhatt, Jeremy Nyitrai, Ning Dai, and Lauren Sassoubre. 2020. "Inactivation of *E. Coli*, *Enterococcus* spp., somatic coliphage, and *Cryptosporidium parvum* in wastewater by peracetic acid (PAA), sodium hypochlorite, and combined PAA-ultraviolet disinfection." *Environmental Science: Water Research & Technology* 6 (1):197. doi: https://doi.org/10.1039/C9EW00837C.

Heaven, S., C. J. Banks, L. N. Pak, and M. K. Rspaev. 2007. "Wastewater reuse in central Asia: Implications for the design of pond systems." *Water Science and Technology* 55 (1-2):85–93.

Islam, Shafi Noor, Sandra Reinstädtler, and Saeid Eslamian. 2015. "Water reuse sustainability in cold climate regions." In *Urban water reuse handbook*, chap 68, 875–886. Taylor and Francis/CRC Group.

Jackson, S., Rodda, N., & Salukazana, L. (2006). "Microbiological assessment of food crops irrigated with domestic greywater." *Water S.A*, 32(5): 700–704.

Jenssen, Petter D. 2004. "Decentralized urban grey water treatment at Klosterenga Oslo." Department of Mathematical Sciences and Technology, *Agricultural University of Norway*, 1432.

Jenssen, P. D. & Vråle, L. (2003). Greywater treatment in combined biofilter/constructed wetlands in cold climate. Ecosan – Closing the Loop. Proc. 2nd Int. Symp. Ecological Sanitation, Lübeck Apr. 7-11. 2003, eds. Werner et al., GTZ: Germany, pp. 875–881, 2003.

Jenssen, Petter D, Lasse Vråle, and Oddvar Lindholm. 2007. "Sustainable wastewater treatment." Proceedings of International Conference on Natural Resources and Environmental Management and Environmental Safety and Health, Kuching, Malaysia.

Kataki, S., S. Chatterjee, M. G. Vairale, S. Sharma, S. K. Dwivedi, and D. K. Gupta. 2021. "Constructed wetland, an eco-technology for wastewater treatment: A review on various aspects of microbial fuel cell integration, low temperature strategies and life cycle impact of the technology." *Renewable and Sustainable Energy Reviews* 148:111261. https://doi.org/10.1016/j.rser.2021.111261.

Khurelbaatar, Ganbaatar, Manfred van Afferden, Christopher M. Sullivan, Christoph Führer, Jamsaran Amgalan, Jöerg Londong, and Roland Arno Müller. 2021. "Wastewater treatment and wood production of willow system in cold climate." *Water* 13 (12):1630.

Kinsley, C., K. Kennedy, and A. Crolla. 2017. "A combined reed bed/freezing bed technology for septage treatment and reuse in cold climate regions." *Water Science and Technology* 76 (2):286–293.

Knutsson, J., and P. Knutsson. 2021. "Water and energy savings from greywater reuse: A modelling scheme using disaggregated consumption data." *International Journal of Energy and Water Resources* 5 (1):13. doi: https://doi.org/10.1007/s42108-020-00096-z.

Lakho, Fida Hussain, Jarne Vergote, Hafiz Ihsan-Ul-Haq Khan, Veerle Depuydt, Teun Depreeuw, Stijn W. H. Van Hulle, and Diederik P. L. Rousseau. 2021. "Total value wall: Full scale demonstration of a green wall for grey water treatment and recycling." *Journal of Environmental Management* 298:113489. https://doi.org/10.1016/j.jenvman.2021.113489.

Lazarova, V., S. Hills, and R. Birks. 2003. "Using recycled water for non-potable, urban uses: A review with particular reference to toilet flushing." *Water Supply* 3 (4):69. doi: https://doi.org/10.2166/ws.2003.0047.

Leung, Ada. 2004. *Advanced studies in Canadian sustainable design*, University of Waterloo.

Mara, Duncan, and Annika Kramer. 2008. "The 2006 WHO guidelines for wastewater and greywater use in agriculture: A practical interpretation." In *Efficient management of wastewater*; Baz, I.A., Otterpohl, R., Wendland, C., Eds.; Springer: Berlin/Heidelberg, Germany. 1–17.

Mohsin, Muhammad, Erik Kaipiainen, Mir Md Abdus Salam, Nikolai Evstishenkov, Nicole Nawrot, Aki Villa, Ewa Wojciechowska, Suvi Kuittinen, and Ari Pappinen. 2021. "Biomass production and removal of nitrogen and phosphorus from processed municipal wastewater by *Salix schwerinii*: A field trial." *Water* 13 (16):2298.

Nagpal, Himanshu, Jan Spriet, Madhu Krishna Murali, and Aonghus McNabola. 2021. "Heat recovery from wastewater—A review of available resource." *Water* 13 (9):1274.

Neal, John. 1996. "Wastewater reuse studies and trials in Canberra." *Desalination* 106 (1):399–405. https://doi.org/10.1016/S0011-9164(96)00135-X.

Nowak, Otto, Peter Enderle, and Petar Varbanov. 2015. "Ways to optimize the energy balance of municipal wastewater systems: Lessons learned from Austrian applications." *Journal of Cleaner Production* 88:125–131. https://doi.org/10.1016/j.jclepro.2014.08.068.

Ottoson. J., 2003. Hygiene aspects of greywater and greywater reuse. Licenciate thesis. Royal Institute of Technology/Swedish Institute for Infectious Disease Control. Royal Institute of Technology (KTH) Department of Land and Water Resources Engineering Swedish Institute for Infectious Disease Control (SMI) Department of Water and Environmental Microbiology, Stockholm.

Samuel, Ojo, Mohd Hafiz Dzarfan Othman, Roziana Kamaludin, Oulavanh Sinsamphanh, Huda Abdullah, Tonni Agustiono Mohd Hafiz Puteh, Tao Kurniawan, Ahmad Fauzi Li, Mukhlis A. Ismail, Juhana Rahman, Tijjani Jaafar, Stanley El-badawy, and Chinedu Mamah. 2022. "Oilfield-produced water treatment using conventional and membrane-based technologies for beneficial reuse: A critical review." *Journal of Environmental Management* 308:114556. https://doi.org/10.1016/j.jenvman.2022.114556.

Scott, Christopher A, Pay Drechsel, Liqa Raschid-Sally, Akiça Bahri, Duncan Mara, Mark Redwood, and Blanca Jiménez. 2009. "Wastewater irrigation and health: Challenges and outlook for mitigating risks in low-income countries." In *Wastewater irrigation and health*, 407–420. Routledge.

Sengupta, Sukalyan, Tabish Nawaz, and Jeffrey Beaudry. 2015. "Nitrogen and phosphorus recovery from wastewater." *Current Pollution Reports* 1 (3):155. doi: https://doi.org/10.1007/s40726-015-0013-1.

Shelef, G., and Azov, Y. (1996). "The coming era of intensive wastewater reuse in the Mediterranean region." *Water Science and Technology*, 33(10), 115–125.

Shen, Chao, Zhuoyu Lei, Yuan Wang, Chenghu Zhang, and Yang Yao. 2018. "A review on the current research and application of wastewater source heat pumps in China." *Thermal Science and Engineering Progress* 6:140–156.

Tierney, J T, R Sullivan, and E P Larkin. 1977. "Persistence of poliovirus 1 in soil and on vegetables grown in soil previously flooded with inoculated sewage sludge or effluent." *Applied and Environmental Microbiology* 33 (1):109. doi: https://doi.org/10.1128/aem.33.1.109-113.1977.

Uddin, Sayed Mohammad Nazim, Zifu Li, Tobias Ulbrich, Heinz-Peter Mang, Jan Franklin Adamowski, and Roman Ryndin. 2016. "Household greywater treatment in water-stressed regions in cold climates using an 'Ice-block Unit': Perspective from the coldest capital in the world." *Journal of Cleaner Production* 133:1312–1317. https://doi.org/10.1016/j.jclepro.2016.06.063.

US Environmental Protection Agency. 1999. "Wastewater Technology Fact Sheet Chlorine Disinfection; EPA 832-F-99-062."

Van Rossum, Tony. 2020. "Water reuse and recycling in Canada — History, current situation and future perspectives." *Water Cycle* 1:98–103. https://doi.org/10.1016/j.watcyc.2020.07.001.

Vassos, Troy. 2014. "Water Reclamation & Reuse: Understanding the promise – and the potential." *Municipal World.*

Vazquez-Montiel, Oscar, Nigel J. Horan, and Duncan D. Mara. 1996. "Management of domestic wastewater for reuse in irrigation." *Water Science and Technology* 33 (10):355–362. https://doi.org/10.1016/0273-1223(96)00438-6.

Wang, H.-C., Cui, D., Han, J.-L., Cheng, H.-Y., Liu, W.-Z., Peng, Y.-Z., Chen, Z.-B., and Wang, A.-J. (2019). "A2O-MBR as an efficient and profitable unconventional water treatment and reuse technology: A practical study in a green building residential community." *Resources, Conservation and Recycling*, 150: 104418.

Wang Mo, Dong Qing Zhang, Jian Wen Dong, and Soon Keat Tan. 2017. "Constructed wetlands for wastewater treatment in cold climate—A review." *Journal of Environmental Sciences* 57:293–311.

Wang Yongli, Jiapu Li, Shuo Wang, Jiale Yang, Chengyuan Qi, Hongzhen Guo, Ximei Liu, and Hongqing Zhang. 2020. "Operational optimization of wastewater reuse integrated energy system." *Energy* 200:117309. https://doi.org/10.1016/j.energy.2020.117309.

Whitlock, Drury, Julian Sandino, Tim Shea, and Peter Burrowes. 2009. "Sustainable Energy Management: Achieving Energy Independence at Wastewater Utilities." Residuals and Biosolids Conference 2009.

WHO (2006a). Guidelines for the Safe Use of Wastewater, Excreta and Greywater—Vol II—Wastewater Use in Agriculture, 3rd ed.; World Health Organization: Geneva, Switzerland, 2006; Volume 2.

WHO (2006b). Guidelines for the Safe Use of Wastewater, Excreta and Greywater—Vol. III. Wastewater and excreta use in aquaculture, 3rd ed.; World Health Organization: Geneva, Switzerland, 2006; Volume 3.

Wu, Shubiao, David Austin, Lin Liu, and Renjie Dong. 2011. "Performance of integrated household constructed wetland for domestic wastewater treatment in rural areas." *Ecological Engineering* 37 (6): 948–954. https://doi.org/10.1016/j.ecoleng.2011.02.002.

Wu, Shubiao, Peter Kuschk, Hans Brix, Jan Vymazal, and Renjie Dong. 2014. "Development of constructed wetlands in performance intensifications for wastewater treatment: A nitrogen and organic matter targeted review." *Water Research* 57:40–55.

Yu, Zita L. T., Anditya Rahardianto, J. R. DeShazo, Michael K. Stenstrom, and Yoram Cohen. 2013. "Critical review: Regulatory incentives and impediments for onsite graywater reuse in the United States." *Water Environment Research* 85 (7):650–662. https://doi.org/10.2175/106143013X13698672321580.

Zhang, L., Lin Mu, Ying Xiong, Beidou Xi, Guowen Li, and Caole Li. 2015. "The development of a natural heating technology for constructed wetlands in cold climates." *Ecological Engineering* 75. doi: https://doi.org/10.1016/j.ecoleng.2014.11.025.

Zhang, Tao, Rongfeng Jiang, and Yaxin Deng. 2017. *Phosphorus recovery by struvite crystallization from livestock wastewater and reuse as fertilizer: A review*: IntechOpen.

Part II

Environmental and Techno-Economic Aspects of Decentralized Systems

6 Environment and Hygiene with Decentralized Treatment

Denisse Serrano-Palacios[a]*, Ana María Rentería-Mexía*[a]*, Lourdes Mariana Díaz-Tenorio*[a]*, Edna Rosalba Meza-Escalante*[a]*, Luis Humberto Álvarez-Valencia*[a]*, Pasiano Rivas-García*[b]*, Gabriela Ulloa-Mercado*[a]*, Pablo Gortáres-Moroyoqui*[a]*, and Luis Alonso Leyva-Soto*[a,c]

[a] Instituto Tecnológico de Sonora, Sonora, México

[b] Departamento de Ingeniería Química, Facultad de Ciencias Químicas, Universidad Autónoma de Nuevo León. Nuevo León, México

[c] Consejo Nacional de Ciencia y Tecnología (CONACYT), Sonora, México

6.1 INTRODUCTION

Water, energy, and nutrients are crucial elements inherent to human beings (Sheikh et al., 2019; Estévez et al., 2022). As far as water is concerned, decentralized wastewater management systems (DWMS) have an enormous impact on multiple aspects of human life, providing methods for recycling and supplying water and nutrients for diverse uses. For example, recovery water from DWMS is reused globally for landscape and agricultural irrigation, industrial use, non-potable domestic uses, groundwater recharge, and recreational uses (Lautze et al., 2014; González et al., 2015).

Centralized systems for water and sanitation services are suitable methods for large-population settlements but can be impractical for many communities, significantly distant or with slight urban growth (Iribarnegaray et al., 2016). Such technologies tend to have lower per capita costs, produce more significant social and environmental gains, and maximize the number of people with access to improved sanitation (Reed, 1996). Thus, governments prioritize large cities while the most remote or smaller communities are unprotected (Bakir, 2001). This means that the coverage rate of sewerage systems in rural areas needs to catch up to urban areas (Hophmayer-Tokich, 2006). On the other hand, conventional centralized systems require highly skilled labor, large amounts of capital due to high costs and complexity of construction, operation, and maintenance, and high and reliable water consumption; they are energy intensive and rank as a significant contributor to greenhouse gas emissions (Xue et al., 2019). Therefore, centralized systems may be less suitable for low-income areas, rural areas with low population density, water-scarce regions, and unreliable water supply systems (Jackson, 1996; Bakir, 2001). A viable alternative in these cases is the DWMS, which allows the treatment of domestic wastewater close to its source of origin, and is the most suitable in rural areas and communities located on the periphery of large human settlements; these sites have limited or no sanitation infrastructure at all (Rodríguez, 2009; De Anda Sánchez, 2017). In addition, these systems minimize the risks to the health of inhabitants from the rural sector, improving their quality of life and the environment (Valencia et al., 2010). Some DWMS advantages are relatively low-tech, low-cost technologies that allow construction and operation by the local community (UNEP/GPA, 2000). One of the main advantages of DWMS is the possibility of reducing potable water demand through on-site or near-site water reuse, depending on the type of community, technical options, and local

DOI: 10.1201/9781003376415-8

environments; favoring the reduction of the effects on the environment and local public health problems related to wastewater (De Anda Sánchez, 2017).

6.2 EFFICIENCY OF DECENTRALIZED TREATMENT METHODS

The wastewater treatment systems in the urban and rural sectors are characterized by being centralized, with complex technologies, and demand high-cost operation and maintenance (Valencia et al., 2010; De Anda Sánchez, 2017). Therefore, in recent years, different types of systems with nonconventional and decentralized features have been proposed (Valencia et al., 2010). Developing countries need help with water supply and wastewater management. The sewage systems do not exist or exist in a rather rudimentary way, and the wastewater generated needs to be treated (Wilderer and Schreff, 2000). This is a challenge not only for developing countries; urban, peri-urban, and rural areas in various parts of the world still require low-cost, reliable, and affordable wastewater treatment services (De Anda Sánchez, 2017).

DWMS consist of collecting, treating, and reusing or disposing of wastewater from houses, isolated communities, or organizations near the point of wastewater generation (Rodríguez, 2009; Suneethi et al., 2015). Although decentralized systems are as easy to build, operate, and maintain, they must carry out essential operation and maintenance activities to remain efficient (Valencia et al., 2010).

A decentralized system can become competitive when satisfying the following conditions: (1) the wastewater treatment systems must be highly effective, robust, easy to operate, and low in costs, and (2) the operation and control of the treatment systems must be by people who have training for this job (Wilderer and Schreff, 2000).

Most technologies currently applied to treat small quantities of wastewater need to meet the requirements (Wilderer and Schreff, 2000). An inappropriate selection of technologies could be due to the non-availability of low-cost technologies to suit different urban areas in developing countries (Sundaravadivel and Vigneswaran, 2001).

Defining appropriate technologies is necessary to address the following aspects: economic, social, and the predominant diseases, risks, and particular technological developments, which translate into sustainability and growth (Rodríguez, 2009). An inappropriate selection of technologies could be due to the non-availability of low-cost technologies to suit different urban areas in developing countries (Sundaravadivel and Vigneswaran, 2001). In addition, little knowledge and field data are available on some basic design, operation, and monitoring parameters due to the relatively recent decentralized wastewater treatment system (DEWATS) implementation (Reynaud and Buckley, 2015).

The main design parameters for DEWATS are wastewater production and organic load per capita. Unfortunately, there is scarce literature available for communities in developing countries, so designers use estimates for plant size (Reynaud and Buckley, 2015).

The best technology for wastewater treatment depends on three essential factors: technical, economic, and environmental (Rodríguez, 2009; Massoud et al., 2009; Capodaglio, 2017). From a technical point of view, the characteristics of the wastewater, the treatment objectives, and the corresponding discharge standards must be considered (Rodríguez, 2009).

Among the causes of failures in decentralized systems are a lack of knowledge about operation and maintenance, intermittent power supply, process instability, plant aging, hydraulic overload, and problems in design and installation (Suneethi et al., 2015).

6.3 POTENTIAL PATHOGEN RISK RELATED TO HYGIENE

Black and kitchen wastewaters are the streams with great organic matter content, but for the presence of feces, black wastewater (BW) presents a microbiota with a particular diversity, distribution, and abundance. Some of them are human pathogenic: viruses, bacteria, protozoans, and helminths (Gyawali, 2018; Jia and Zhang, 2020); some examples are shown in Table 6.1.

TABLE 6.1
Pathogenic Microorganisms and Organisms Commonly Detected in Wastewater

Pathogen Type	Genus, Species, or Family	
Bacteria	*Brucella* spp.	
	*Campylobacter jejuni**	
	Clostridium perfringens, *C. tetani*	
	Enteroinvasive or enteropathogenic or enterotoxigenic or enterohemorrhagic *Escherichia coli*	
	Francisella tularensis	
	Leptospira interrogans	
	Mycobacterium tuberculosis	
	Nocardia spp.	
	Salmonella spp.*, *Salmonella paratyphi*, *S. typhy*	
	Shigella spp.*	
	Vibrio cholerae, *V. parahaemolyticus*	
	Yersinia enterocolitica	
Viruses	Coronavirus	Norovirus
	Enterovirus	Hepatitis virus
	Rotavirus	Adenovirus
Protozoa	*Cryptosporidium*	
	Giardia	
	Entamoeba histolytica	
Helminths	*Ascaris* spp.	
	Ancylostoma duodenale	
	Necator americanus	
	Trichuris trichiura	
	Enterobius vermicularis	
	Strongyloides stercoralis	
	Toxocara spp.	
	Taenia spp.	
	Hymenolepis spp.	
	Echinococcus spp.	
	Dicrocoelium dendriticum	
	Fasciola spp.	
	Schistosoma spp.	

* Could be considered as priority pathogens if strains are resistant to fluoroquinolone.
Source: Gyawali (2018); Jia and Zhang (2020).

Although the effluents from DWMS can be used for street washing or agricultural and landscape irrigation, it is essential to ensure that they do not cause any infection or disease from pathogen microorganisms or parasites. Hence, a study about the exposure risk for direct or indirect interaction needs to be evaluated, considering factors such as possible products or structures exposed to effluent following contact with people. Other factors are effluent's proximity to human settlement and their time-frequency exposure. Because of this, health and hygiene are part of the evaluation criteria for choosing the most suitable technology in DWMS (Lourenço and Nunes, 2020). Furthermore, in 2017, the World Health Organization (WHO) published a priority pathogens list for research and development of new antibiotics; in Table 6.1, *Campylobacter jejuni*, *Salmonella*, and *Shigella* spp. could be considered if strains were resistant to fluoroquinolone (WHO, 2022a), therefore their monitoring is for sanitary purpose.

Public health and ecological factors are the main worries for designing a DWMS; identifying human pathogens is necessary for wastewater streams and effluent to ensure health in a given location. Not all of them are equally important, but each one could cause damage either to a single person or the whole community. In India, regulations for the quality of effluents set limits for the count of fecal coliforms. Particular interest is shown in multi-drug resistant pathogens that can be found in treated wastewater; they may cause disease, and the treatment is aggressive, long, and expensive. If the DWMS design contemplates reuse for vegetable irrigation, a chlorination step is suitable (Starkl et al., 2022). Physical, chemical, and biochemical tools were recently developed to detect, identify, and quantify pathogens. In recent decades, DNA and RNA molecular tools have been extensively used for the specific and sensitive detection of pathogens in wastewater. At least two methods have been applied to confirm the results.

Severe acute respiratory syndrome (SARS), caused by a coronavirus (SARS-CoV-1), was an enormous public health concern. It is an enveloped virus, and it is persistent in raw sewage at 20°C for two days and 14 days at 4°C; studies have shown that the viability of SARS coronaviruses has not been observed in the effluent in treated wastewater. Also, it has been reported that it can be inactivated using chlorine (> 0.5 mg/L) or chlorine dioxide (> 2.19 mg/L). Mechanisms of removal of viruses include adsorption of viruses on larger aggregated particles, which are then separated by sedimentation. After that, delicate operations like retention by a membrane or biofilm layers, predation, and enzymatic breakdown in membrane bioreactors are used, moreover the traditional tools like UV exposure, chlorination, and ozonation (Corpuz et al., 2020).

6.4 ENVIRONMENTAL IMPACTS OF DECENTRALIZED TREATMENT

The COVID-19 pandemic has underscored the importance of ensuring safe water, sanitation and waste management, and hygienic conditions for preventing and protecting human health during infectious disease outbreaks. However, 2.3 billion people, or 29% of the global population, lacked a primary handwashing facility with water and soap at home in 2020 (WHO, 2021). Access to safe water, sanitation, and hygiene is vital for health and for social and economic progress. Although billions of people in the world lack access to safely managed sanitation services (with excreta safely disposed of on-site or treated off-site), about 46% of the world's population lacks basic sanitation, and sanitation coverage is commonly much lower in rural areas than in urban areas (WHO, 2022b).

Worldwide, significant development has been made in wastewater treatment for urban areas compared to rural areas, which lag far behind. However, wastewater treatment plants represent one of the significant investments due to high capital costs in addition to operation and maintenance costs. Furthermore, restricted local budgets, lack of local expertise, and lack of funding result in the inadequate operation of wastewater treatment plants in developing countries (Paraskevas et al., 2002).

Moreover, decentralized systems can be designed for a specific site, thus overcoming the problems associated with site conditions such as high groundwater tables, impervious soils, shallow bedrock, and limestone formations. Furthermore, decentralized systems allow for flexibility in management, and a series of processes can be combined to meet treatment goals and address environmental and public health protection requirements (Massoud et al., 2009).

Protecting public health (by providing a clean environment and breaking the cycle of disease) and the environment (by avoiding the adverse effects of excessive pollutants discharged into the environment) are the principal concerns in implementing DWMS. Sustainable decentralized sanitation focuses on the on-site treatment of wastewater and local recycling and reusing resources contained in domestic wastewater. Their multiple products are responsible for reduced potable water demand due to on-site or near-site water reuse; bio-energy (biogas produced by anaerobic system from the transformation of organic material); and digestate streams containing nutrients, mainly nitrogen and phosphorus, that are considered biofertilizers for horticultural or ornamental crops in green areas and agricultural lands that surround the facilities (Malila et al., 2019). As an additional advantage, decentralized systems are generally compact, with highly flexible operating conditions and reduced aesthetic impact; however, other local impacts (i.e., odors, traffic) should be considered (Capodaglio, 2017).

In this sense, if water reuse is to gain acceptance, DWMS must create a paradigm shift from effluent disposal (wastewater is a problem) to water reuse (wastewater is a resource). Furthermore, it is essential that the term "reuse," which connotes a resource, be used in place of "disposal," which connotes a problem (Tchobanoglous et al., 2004). Altogether, these benefits contribute to the favorable environmental profile of DWMS, countersigning that these technologies align with the circular economy's philosophy and guidelines (Estévez et al., 2022).

Sustainable resource use in urban waters includes water resources, nutrient management, material use, energy use and greenhouse gas emissions, and waste generation. There is the potential to utilize treated effluent locally, for example, in toilet flushing, washing clothes, cleaning floors, watering lawns and gardens, or local agriculture. In addition, DWMS do not have the potential to cause a severe impact when systems fail, as the area affected by any failure will be limited. In the case of decentralized wastewater specifically, the volumes of effluent released to aqueous environments have less cumulative impacts on parameters such as diluted oxygen, chemical oxygen demand, and biochemical oxygen demand (Singh et al., 2019). Therefore, decentralization could contribute to environmental sustainability and reverse the loss of environmental resources, improving system resiliency, efficiency, lost or diminished environmental functions, and resource recovery (Novotny and Brown, 2007).

Finally, it should be pointed out that decentralization could minimize environmental impacts and public health risks in developing countries' urban, peri-urban, and rural areas. However, some of the following considerations must be achieved: (1) better distribution of sewerage investment costs by reducing distances, diameters, and, sometimes, pumping stations; (2) the possibility of extending wastewater treatment coverage with efficiencies that comply with current discharge regulations (depending on the technology used); and (3) the potential for reuse of effluent for agriculture and other urban activities implies, depending on the case, less pressure on water resources, lower demand for high-quality water, higher rates of return on investment, lower fertilizer consumption, and higher nutrient use and energy recovery.

6.5 IMPACT ON HUMAN LIFE

DWMS is a sustainable initiative to reduce the impact on health risk, quality of life, and the environment. For a system to be environmentally sustainable, it must ensure environmental protection; conservation of resources allows the reuse of water and recycling of nutrients and resources (Capodaglio, 2017). Accordingly, one of the main objectives of wastewater management systems is to provide water quality and ecosystem protection by avoiding the adverse effects of excessive pollutants discharged into the environment. There is a water crisis worldwide, and among the causes are its scarcity and quality due to pollution and eutrophication (World Water Assessment Programme, 2020).

Referring to cost, DWMS provide a long-term and cost-effective option for treating residual effluents with multiple advantages for communities where centralized governance methods are challenging to apply. First, although these technologies are often perceived as too costly (Gómez-Román et al., 2021), DWMS represent a small conveyance network since the effluent treatment, disposal, or reuse of the effluent is close to the source where wastewater is produced and collected. Second, DWMS are used with local treatment methods at home, in neighborhoods, or in small communities, thus reducing costs for transporting effluents (Lens et al., 2005). Finally, the local recovery of the treated water and nutrients allow them to be recycled into the same community, achieving locally sustainable costs and promoting the circular economy (Roefs et al., 2017).

For human health, having quality water is vital. Therefore, reducing pathogens from domestic wastewater is a high priority for environmental and public health protection (Wang et al., 2021). However, although the availability of quality water is a fundamental human right, only 2.9 billion people (approx. 39%) had a suitable sanitation facility, and 1.9 billion people (approx. 27%) used private sanitation services (WHO and UNICEF 2015). Besides, around 2.3 billion people still miss a proper sanitation facility (WHO, 2017). Among the main objectives of wastewater management systems are to protect and promote human health by breaking the cycle of disease (Capodaglio, 2017). DWMS aim to provide proper sanitation methods, providing a good-quality pathogen-reduced water supply for non-potable uses (Bajpai et al., 2019). The most effective removal mechanisms of bacterial and viral pathogens in DWMS may vary depending on numerous factors such as the configuration of the treatment units, filtration materials, hydraulic loadings, seasonal variations, operational conditions (such as hydraulic retention time, flow pattern, recirculation, and aeration), and other environmental factors (such as temperature and pH) (Wang et al., 2021).

Within types of DWMS, and depending on the conditions of each process, filtration-based treatment systems and constructed wetlands have resulted in different removal rates of pathogens. For example, among the physical steps within the DWMS, pathogen removal of large-particle bacteria (>80 μm), helminth ova, and *Giardia* and *Cryptosporidium* as fecal bacteria have been reported in sedimentation (Boutilier et al., 2009; Gikas and Tsihrintzis, 2012).

On the other hand, the effectiveness of filtration mainly depends on the characteristics of the pathogen and filtration media, including the type, texture, and size (Morató et al., 2014). Efficient bacteria (4.85–6.8 log10 CFU/100 mL) and protozoa (2 log10 CFU/100 mL) removal have been observed through the filtration process (Redder et al., 2010), with inefficient removal for viruses due to their small sizes (Azadpour-Keeley et al., 2003).

Within chemical removal mechanisms, adsorption becomes the leading mechanism for pathogen removal in which several factors affect their efficiencies, such as physical (e.g., media types, organic matter content, and temperature), chemical (e.g., pH and ionic strength), and microbiological (e.g., cell types and cell surface characteristics) (Sasidharan et al., 2016). Removal of viruses and bacteria through adsorption to soils is strongly affected by the type and strain of viruses and bacteria because their surface characteristics vary significantly, as reported in a study with fecal bacteria (*E. faecalis* and *E. coli*) (Chen and Walker, 2012).

Another effective mechanism for pathogen removal, mainly in open-water treatment wetlands (e.g., free water surface flow and waste stabilization ponds in constructed wetlands), is UV radiation by sunlight (Davies-Colley et al., 1999). The effects of UV radiation depend on the amount of sunlight the systems have access to, for example, in wetlands (Nguyen et al., 2015), during enterococci and fecal-specific bacteriophage inactivation by the three regions of sunlight (Silverman et al., 2013), and the inactivation of *E. coli*, poliovirus, adenovirus, PRD1, and MS2 bacteriophages by UV-B (Kadir and Nelson, 2014). Other factors affecting pathogen removal by UV radiation are the different organisms' survivability, vegetation density, water depth, and quality (Wu et al., 2016; Shingare et al., 2019). However, the effect of UV radiation by sunlight is negligible in subsurface flow-constructed wetlands and most filtration-based DWMS.

Another mechanism in removing bacterial pathogens from domestic wastewater is natural die-off, controlled by predation, starvation, and exposure to unfavorable factors such as high temperature, low moisture content, low pH, and low organic matter content (Zaleski et al., 2005). Among the environmental conditions affecting the natural die-off of pathogens, seasonal changes are significant since factors such as temperature, pH, composition, and water turbidity are affected (García et al., 2008). From them, temperature is the main factor for the survival of viruses in the environment, since higher temperatures can damage viral DNA or RNA, increase predation and the natural die-off rates of bacteria (Haig et al., 2015), as reported in studies with *Pseudomonas* spp. (Vandenhove et al., 1991) and enteric bacteria (Sjogren, 1994).

Bacterial and viral removal can also be affected by pH. The low pH in some DWMS can result in faster natural die-off rates of bacteria, thus increasing the bacteria removal efficacy, as reported in studies with coliform bacteria (Pundsack et al., 2001). Moreover, turbidity, organic and nutrient content, and wastewater quality also influence the removal of pathogens in DWMS (Díaz et al., 2010; Jenkins et al., 2011). Although indicator pathogens, such as total coliforms, fecal coliforms, *E. coli*, and fecal-specific bacteriophage presented here, have been widely used to represent the water quality (Gagliardi and Karns, 2000), these pathogens are not able to provide information about the existence of other specific pathogens, such as SARS-CoV-2.

Concerning different species, the survivability and morphology of pathogens are some other factors that impact the removal process of pathogens in the DWMS. Such is the case of the longer survival times reported in *Salmonella* and *Streptococcus*, compared *to E. coli* and *Shigella* (Winfield and Groisman, 2003), or the different survival times in studies with *C. perfringens* and *C. parvum* (Hijnen et al., 2007). The longer survivability of some bacteria is attributed to their more remarkable ability to compete for nutrients with indigenous microorganisms; enhanced adsorption to the soil surface; or higher resistance to physical, chemical, or biological stressors in the environment (Stevik et al., 2004).

According to all aspects we have analyzed in this section, the impact of DWMS on human life is influential in the sense that DWMS are significant sustainable technologies for communities. Furthermore, these treatments are economically affordable, environmentally protective, technically and institutionally consistent, and socially acceptable for water recovery (Capodaglio, 2017) as a non-renewable resource vital for human life.

6.6 RISK MANAGEMENT USING DECENTRALIZED TREATMENT

Risk can be defined as the potential of positive and negative objective deviations. Risk management is a tool to care for the safety engineering of the technical unit and the protection of material assets, ecosystems, or human beings. It is an entire protocol to ensure the wastewater system achievement, offering the possibility to cope with the risks step-by-step. The risk management process consists of three steps: risk identification, risk analysis/risk assessment, and risk handling; all are related (Wagner and Strube, 2005).

Wastewater treatment-relevant risks are divided into four types: market, legal, operational, and financial. As we mentioned previously, DWMS has no regulations and is designed for a specific area; thus, the last two risks are more manageable. The risk identification aim is to view from the angle of possible impacts all activities and factors that involve defects and damages to the process and the environment (Wagner and Strube, 2005). Risk assessment estimates the probable impact of a chemical, physical, microbiological, or psychosocial hazard on a specified human population or ecosystem under specific conditions at a given timeframe (Cookey, 2013). Figure 6.1 shows a framework for integrated risk assessment of DWMS; all the factors are evaluated in an orderly way, concerning two-way communication between the risk manager and interested groups. The risk analysis step measures the possible damage level and the occurrence probability; the assessment is based on a subjective but realistic estimation. Finally, risk handling refers to coping with the risks to avoid or mitigate the risk that has been identified, analyzed, and assessed. There are two ways to face it: cause-oriented and effect-oriented strategies; the first takes preventive actions to avoid damage or at least reduces the probability that it occurs; the second is related to the handling when the damage is presented, either to slight damage or cover the financial issues (Wagner and Strube, 2005).

Since DWMS is used to treat wastewater for specific small towns, several types exist. However, there is not vast information about the performance, control, and social perspectives concerning them. Iribarnegaray et al. (2018) analyzed 24 informal DWMS in Lerma Valley, in the province of Salta in Argentina. They divided the DWMS into three groups: septic tank (ST), ST combined with UASB reactor (UASBR), and UASBR with a final filtration step (UASBR-F). The groups presented statistical differences ($p < 0.05$) in effluent characteristics, such as chemical oxygen

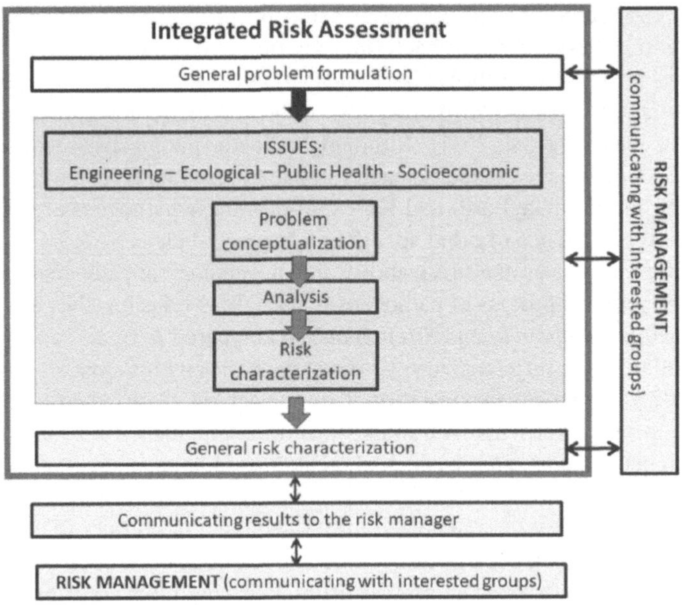

FIGURE 6.1 Integrated risk assessment framework of DWMS. (Adapted from Jones et al., 2000.)

demand and total and thermotolerant coliforms. In all cases, this last bacteria group did not comply with local environmental legislation for the lack of a final disinfection procedure. ST was the least efficient treatment. Meanwhile, for UASBR and UASBR-F treatments, no statistical differences were observed. In this case, a re-evaluation is needed to prepare a proposal where coliform standards comply with local regulations, to reduce the potential risk for public health and ecological issues.

For public health and ecological reasons, risk assessment protocols have to be dynamic and constantly improved upon. Besides biological and microbiological issues, other wastewater and effluent characteristics, such as antibiotic resistance genes (ARG), should be monitored. Zhao et al. (2022) aims to understand the risks to human health and the ecosystem where the effluent passes; they can affect livestock breeding or fish farming in an infectious outbreak, causing a low growth rate and an increment in mortality rate, bringing seafood shortages, price increases, and food quality loss as consequences. Influents and effluents from developed rural areas in eastern China DWMS were analyzed, and a total of 825 bacterial and 19 viral genotypes were identified in wastewater. Table 6.2 shows the leading bacterial genus and viral family genotypes. Just *Pseudomonas* and *Mycobacterium* spp. are listed for an urgent new therapy by the World Health Organization (Zhao et al., 2022).

At waste streams, the dominant ARGs found include bacitracin (27.58%), tetracycline (8.7%), sulfonamide (6.2%), and penicillin related (3.9%), with relative abundances of 27.8%, 8.7%, 6.2%, and 3.9%, respectively. On the other hand, dominant ARGs from effluents sampled included bacitracin (36.8%), tetracycline (4.3%), sulfonamide (7.4%), and penicillin (4.1%) classes. The authors discuss that results differ from typical municipal central wastewater systems, where dominant ARGs mainly belong to beta-lactams, quinolone, aminoglycoside, tetracycline, and sulfonamide.

Bacteria and viruses, especially phages, contribute to ARG spreading. It was found that some ARGs correlated with bacteria and phages. Therefore, risk analysis from this factor must be done, mainly to the risk of spreading ARGs through phages.

The development of risk management for DWMS are mandatory to identify risks for a well-preparation response when a risk appears. Risk management is the end step of risk-based decision-making, which should systematically result in technological improvement and mitigation and

TABLE 6.2
Main Genotypes of Bacteria Genus and Virus Families Found in Waste Streams and Effluents from Eastern China DWMS

Bacteria Genotypes (%)		Virus Genotypes	
Waste Stream	**Effluent**	**Waste Stream**	**Effluent**
Flavobacterium (14.3%)	*Acidovorax* (14.5%)	Myoviridae	
Acidovorax (10.2%)	*Mycobacterium** (12.4%)	Siphoviridae	
Arcobacter (10.1%)	*Flavobacterium* (9.8%)	Podoviridae	
Limnohabitans (8.6%)	*Pseudomonas** (8.5%)		
*Pseudomonas** (7.5%)			

* These genera are listed in urgent new therapy by the WHO. Data obtained from: Zhao et al. (2022).

remediation of failed technologies for the equilibrium among engineering, ecological, socioeconomic, and public health factors.

6.7 HEALTH AND ENVIRONMENTAL IMPACT OF DECENTRALIZED SYSTEM VS. CENTRALIZED SYSTEM TREATMENT

DWMS are designed to operate on a small scale, reduce environmental and public health impacts, and increase the ultimate reuse of wastewater, depending on the type of community, technical facilities, and local demands. When decentralized systems are used effectively, they promote the return of treated wastewater to any water bodies nearby of the discharge. In addition, decentralized systems can be installed on an as-needed basis, thus avoiding the costly implementation of centralized treatment systems. Unlike centralized wastewater treatment systems, decentralized systems are preferable for communities with low population density (Massoud et al., 2009).

The advantages of wastewater management concerning the characteristics of decentralized treatment systems are depicted in Figure 6.2.

Regardless of the method used, wastewater management is essential for several reasons, one of which is to protect public health and the well-being of the communities (Hophmayer-Tokich, 2006). Sustainable development demands conserving energy, protecting environmental integrity, and promoting public health. Therefore, it is critical to assess the health risks of wastewater treatments. One of the emerging quality markers for these treatments is the analysis of health impacts (Schoen et al., 2017). Among indicators to evaluate health impact, life cycle assessment (LCA) and quantitative microbial risk assessment are excellent approaches. LCA is a systematic evaluation approach that aims to quantify the environmental health impacts of chemical releases from a process, product, or service, according to ISO 14040:2006. LCA includes chemical health risks caused by treatment technologies and their supply chain processes, such as producing the required energy and coagulant. A few recent LCA studies have begun to include human health impacts on urban water service (Lane et al., 2015).

LCA consists of four phases: goal and scope definition, inventory analysis, impact assessment, and interpretation. Characterization factors for life cycle health impacts utilize the USEtox model. USEtox is a model officially endorsed by the United Nations Environment Program/Society of Environmental Toxicology and Chemistry (UNEP/SETAC) for comparative assessment of human health impacts of chemicals in conjunction with LCA modeling. USEtox's characterization factors are based on a combination of multimedia environmental fate, multi-pathway exposure assessment, and toxicity assessment models, and are often expressed in cases/kg of released chemicals. Among the factors of LCA are life cycle cancer and life cycle non-cancer. For example, the life cycle impact

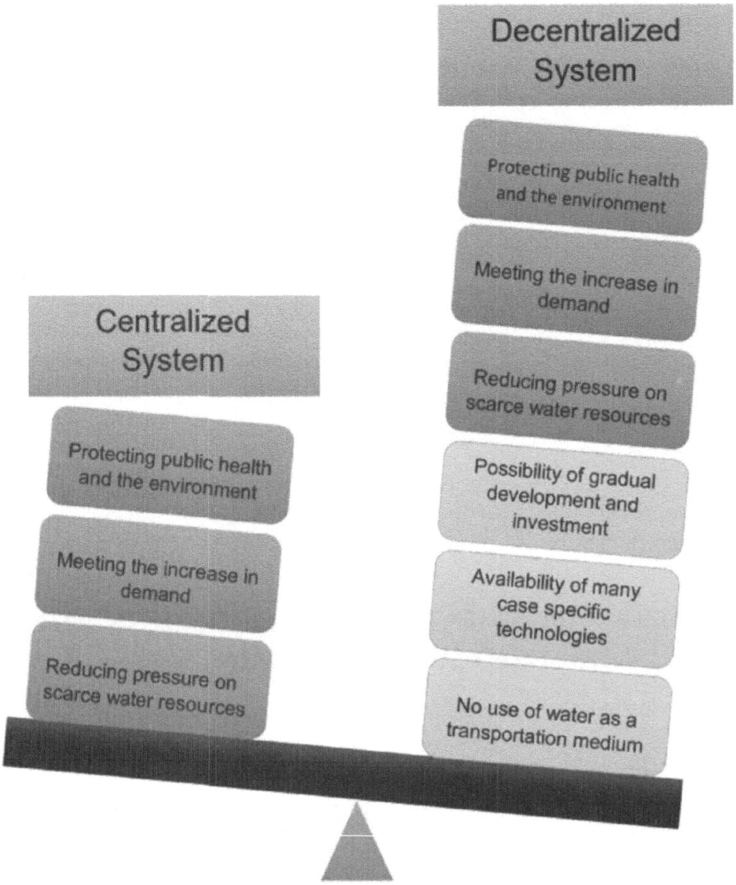

FIGURE 6.2 Advantages of centralized versus decentralized system characteristics. (Adapted from Massoud et al., 2009.)

of a water service system is commonly estimated as the sum of products divided by amounts of released chemicals and their cancer characterization factors. Similarly, a water service system's life cycle non-cancer impact is equal to the sum of products divided by the amounts of released chemicals and their non-cancer characterization factors (Romeiko, 2020).

There are several studies regarding LCA. Romeiko (2020) compared the health impacts of three representative systems by integrating life cycle and microbial risk assessment approaches. Findings indicated that the decentralized system options, such as on-site septic tanks and composting or urine-diverting toilets, presented much lower life cycle cancer and non-cancer impacts and lower microbial risks than the centralized system. Besides, life cycle cancer and non-cancer impacts contributed to approximately 95% of total health impacts, while microbial risks were associated with the remaining 5%. Conclusions from this study included that reducing the energy use of wastewater treatment and water distribution is effective in mitigating total health damages of the centralized system, while reducing the energy use of water treatment is effective in mitigating total health damages of the decentralized systems.

On the other hand, currently existing and projected water scarcity and related factors have forced the water community to focus on the need for water reuse. Moreover, decentralized infrastructure could significantly minimize the risks of drought and extreme events in different locations. It is, therefore, a more climate-resilient option, as well as a lower vulnerability to natural disasters, power outages, and acts of vandalism (Capodaglio, 2017).

Despite mentioning many advantages of the decentralized system over the centralized system, decision-makers must apply indicators such as LCA to evaluate the environmental effects of a decentralized system in a community. The environmental impact case study described next is based on two systems located in Northern California (Shehabi et al., 2012). The decentralized community-scale system provides wastewater treatment to the 47-lot suburban subdivision of Stonehurst in Martinez, California, USA. The system was designed to treat nearly 20 million L of liquid wastewater annually and operates at total capacity. Each parcel includes a 5.7 m^3 concrete septic tank. The centralized system serves approximately half a million residents in an area of 200 km^2 and includes conventional liquid and sludge treatment process streams. Liquid wastes are treated with primary sedimentation, pure oxygen-activated sludge, secondary clarification, and disinfection. Solids are anaerobically digested and dewatered prior to disposal. Figure 6.3 presents life-cycle energy and emissions of greenhouse gas and criteria pollutants for both the centralized and decentralized wastewater treatment systems separated by activity; the results show that the decentralized system

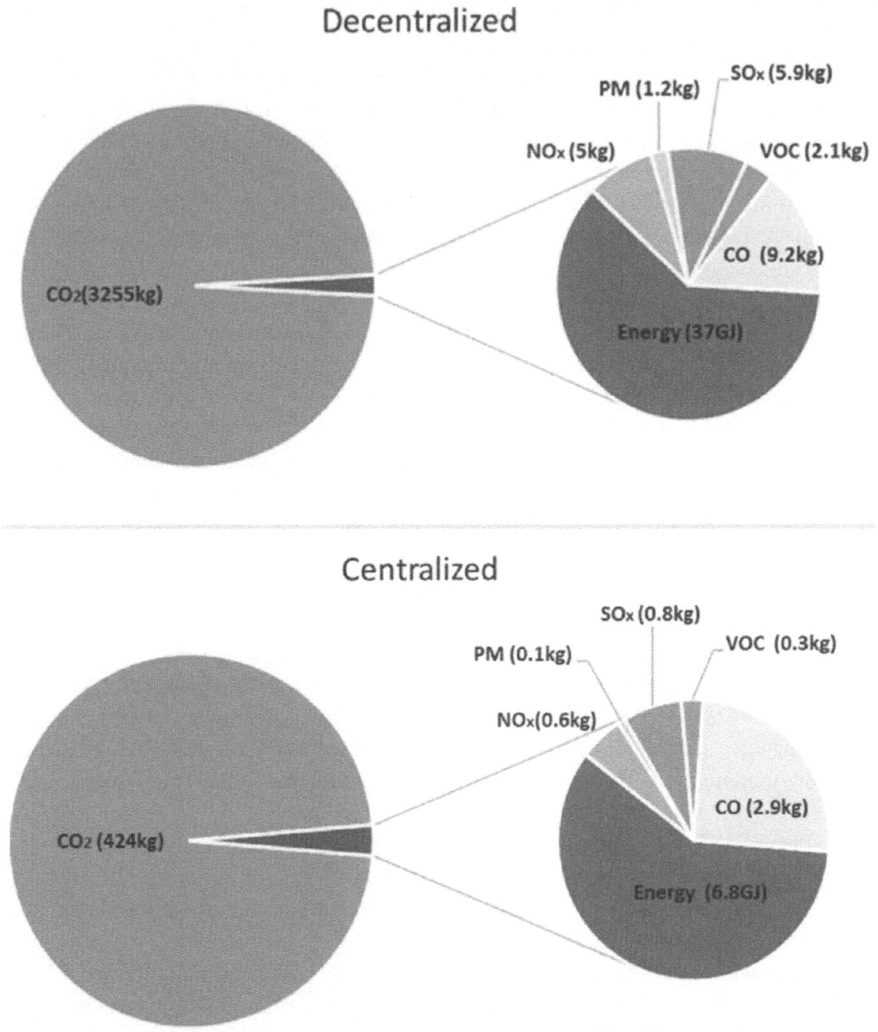

FIGURE 6.3 Case study life-cycle energy use and air emissions by activity. (Key: CO: carbon monoxide; CO_2(e): carbon dioxide equivalent; NOx: nitrogen oxides; PM: particulate matter; SOx: sulfur oxides; VOC: volatile organic compounds.)

used about 37 GJ of primary energy per million liters of treated wastewater, while the centralized system used about 6.8 GJ for the same functional unit. After normalizing for differences in volumes of waste treated, company records indicate that the decentralized system requires seven times more electricity than the centralized system; when comparing greenhouse gas emissions, expressed in units of carbon dioxide equivalent (CO_2), the difference between the two systems increases, which is mainly due to direct methane losses from decentralized septic tanks (Shehabi et al., 2012).

Methods are limited, and previous case studies have shown that life-cycle energy impacts do not necessarily align with other attributes such as public need, ecosystem conservation, or aesthetics (Brix, 1999). Therefore, decision-makers must evaluate proposed decentralized designs through LCA indicators to minimize environmental burdens and qualitative considerations of future wastewater treatment development.

6.8 CONCLUSIONS AND PERSPECTIVES

According to the information presented in this chapter, it is clear that developing new technologies on DWMS is necessary to contribute to the functionality of on-site or near-site water reuse management, improving construction, operation, monitoring, and flexibility/adaptability. They are a viable and necessary alternative for treating domestic sewage to protect human health; minimize environmental impacts; facilitate energy and resource recoveries with local methods for the community; and provide inhabitants of urban, peri-urban, and rural areas access to basic sanitation.

DWMS are aimed to provide accessible treatment services and include wastewater collection, treatment, and reuse or disposal methods. DWMS are mainly used in hilly rural areas, either in single or cluster residences; industrial, business, or institutional facilities; and communities where the usual centralized system cannot be installed due to physical or technical issues (e.g., topography or accessibility to a pumping station network).

As an example of new technologies, updated decentralized installations include a remote control, which implies no need to have trained personnel facilitating operation and maintenance. Moreover, to protect public health, it is vital to minimize the risk of exposure to pathogens and hazardous substances that could affect public health. Furthermore, since DWMS are a long-term and cost-effective option for treating residual effluents with multiple advantages for communities, the environment, and natural resources such as energy, water reuse, and nutrient recovery, are priorities for the circular economy. Finally, financial and economic issues and sociocultural aspects should be considered to achieve equal status, perception, impact, and acceptance of centralized treatment.

Governments prioritize large cities, while the most remote or smaller communities are unprotected. This means that the coverage rate of sewerage systems in rural areas needs to catch up to urban ones. On the other hand, conventional centralized systems require highly skilled labor, large amounts of capital due to high costs and complexity of construction, operation, and maintenance, and high and reliable water consumption. Centralized systems are energy-intensive and rank as a significant contributor to greenhouse gas emissions. Therefore, centralized systems may be less suitable for low-income areas, rural areas with low population density, water-scarce regions, and unreliable water supply systems.

A viable alternative in these cases is the DWMS, which allows domestic wastewater treatment close to its origin source. It is the most suitable in rural areas and communities located on the periphery of large human settlements; these sites have limited or no sanitation infrastructure. In addition, these systems minimize the risks to the health of inhabitants from the rural sector, improving their quality of life and the environment. Some DWMS advantages are relatively low-tech, low-cost technologies that allow construction and operation by the local community. One of the main advantages of DWMS is the possibility of reducing potable water demand through on-site or near-site water reuse, depending on the type of community, technical options, and local environments, favoring the reduction of the effects on the environment and local public health problems related to wastewater.

REFERENCES

Azadpour-Keeley, A., B. R. Faulkner, and J. Chen. 2003. Movement and longevity of viruses in the subsurface. Washington, DC, USA: Environmental Protection Agency, National Risk Management Research Laboratory. Available online: https://cfpub.epa.gov/si/si_public_record_report.cfm?Lab=NRMRL&dirEntryId=64225

Bajpai, M., S. S. Katoch, and N. K. Chaturvedi. 2019. Comparative study on decentralized treatment technologies for sewage and greywater reuse–a review. *Water Science and Technology* 80:2091–2106. https://doi.org/10.2166/wst.2020.039

Bakir, H. A. 2001. Sustainable wastewater management for small communities in the middle East and North Africa. *Journal of Environmental Management* 61:319–328. https://doi.org/10.1006/jema.2000.0414

Boutilier, L., R. Jamieson, R. Gordon, C. Lake, and W. Hart. 2009. Adsorption, sedimentation, and inactivation of *E. coli* within waste water treatment wetlands. *Water Research* 43:4370–4380. https://doi.org/10.1016/j.watres.2009.06.039

Brix, H. 1999. How 'green' are aquaculture, constructed wetlands, and conventional wastewater treatment systems? *Water Science and Technology* 40(3):45–50. https://doi.org/10.1016/S0273-1223(99)00418-7

Capodaglio, A. G. 2017. Integrated, decentralized wastewater management for resource recovery in rural and peri-urban areas. *Resources* 6(2):1–20. https://doi.org/10.3390/resources6020022

Chen, G., and S. L. Walker. 2012. Fecal indicator bacteria transport and deposition in saturated and unsaturated porous media. *Environmental Science & Technology* 46:8782–8790. https://doi.org/10.1021/es301378q

Cookey, P. 2013. Risks assessment of decentralized wastewater treatment systems in Port Harcourt City, Nigeria. Master Science diss. UNESCO-IHE- Institute for Water Education Environmental Science and Technology Delft, the Netherlands. Asian Institute of Technology Sch. of Env. Resources & Development Bangkok, Thailand. Available on line: https://www.researchgate.net/publication/327701372_risks_assessment_of_decentralized_wastewater_treatment_systems_in_port_harcourt_city_nigeria (accessed October 4, 2022).

Corpuz, M.V.A, A. Buonerba, G. Vigliotta, T. Zarra, F. Ballesteros, P. Campiglia, V. Belgiorno, G. Korshin, and V. Naddeo. 2020. Viruses in wastewater: Occurrence, abundance and detection methods. *Science of the Total Environment* 745: 140910. https://doi.org/10.1016/j.scitotenv.2020.140910

Davies-Colley, R. J., A. M. Donnison, D. J. Speed, C. M. Ross, and J. W. Nagels. 1999. Inactivation of faecal indicator microorganisms in waste stabilization ponds: Interactions of environmental factors with sunlight. *Water Research* 33:1220–1230. https://doi.org/10.1016/S0043-1354(98)00321-2

De Anda Sánchez, J. 2017. Saneamiento descentralizado y reutilización sustentable de las aguas residuales municipales en México. *Sociedad y ambiente* 14: 119–143. Available online: http://www.scielo.org.mx/scielo.php?script=sci_arttext&pid=S2007-65762017000200119&lng=es&nrm=iso

Díaz, F. J., A. T. O'Geen, and R. A. Dahlgren. 2010. Efficacy of constructed wetlands for removal of bacterial contamination from agricultural return flows. *Agricultural Water Management* 97:1813–1821. https://doi.org/10.1016/j.agwat.2010.06.015

Estévez, S., S. González-García, G. Feijoo, and M.T. Moreira. 2022. How decentralized treatment can contribute to the symbiosis between environmental protection and resource recovery. *Sci ence of the Total Environment* 812: 151485. https://doi.org/10.1016/j.scitotenv.2021.151485.

Gagliardi, J. V., and J. S. Karns. 2000. Leaching of *Escherichia coli* O157:H7 in diverse soils under various agricultural management practices. *Applied and Environmental Microbiology* 66:877–883. https://doi.org/10.1128/AEM.66.3.877-883.2000

García, M., F. Soto, J.M. González, and E. Bécares. 2008. A comparison of bacterial removal efficiencies in constructed wetlands and algae-based systems. *Ecological Engineering* 32:238–243. https://doi.org/10.1016/j.ecoleng.2007.11.012

Gikas, G. D., and V. A. Tsihrintzis. 2012. A small-size vertical flow constructed wetland for on-site treatment of household wastewater. *Ecological Engineering* 44:337–343. https://doi.org/10.1016/j.ecoleng.2012.04.016

Gómez-Román, C., J. M. Sabucedo, M. Alzate, and B. Medina. 2021. Environmental concern priming and social acceptance of sustainable technologies: The case of decentralized wastewater treatment systems. *Frontiers in Psychology* 12. https://doi.org/10.3389/fpsyg.2021.647406

González, O., B. Bayarri, J. Aceña, S. Pérez, and D. Barceló. 2015. Treatment Technologies for Wastewater Reuse: Fate of Contaminants of Emerging Concern. In *Advanced Treatment Technologies for Urban Wastewater Reuse*, ed. D. Fatta-Kassinos, D. D. Dionysiou, and K. Kümmerer, 5–37. Cham: Springer. https://doi.org/10.1007/698_2015_363

Gyawali, P. 2018. Infectious helminth ova in wastewater and sludge: A review on public health issues and current quantification practices. *Water Science & Technology* 77(4): 1048–1061. https://doi.org/10.2166/wst.2017.619.

Haig, S. J., M. Schirmer, R. D'Amore, J. Gibbs, R. L. Davies, G. Collins, and C Quince. 2015. Stable-isotope probing and metagenomics reveal predation by protozoa drives *E. coli* removal in slow sand filters. *ISME J* 9(4):797–808. https://doi.org/10.1038/ismej.2014.175

Hijnen, W. A. M., Y. J. Dullemont, J. F. Schijven, A. J. Hanzens-Brouwer, M. Rosielle, and G. Medema. 2007. Removal and fate of *Cryptosporidium parvum*, *Clostridium perfringens* and small-sized centric diatoms (*Stephanodiscus hantzschii*) in slow sand filters. *Water Research* 41:2151–2162. https://doi.org/10.1016/j.watres.2007.01.056

Hophmayer-Tokich, S. 2006. *Wastewater Management Strategy: Centralized v. Decentralized Technologies for Small Communities*. Enschede: Center for Clean Technology and Environmental Policy. Available online: https://ris.utwente.nl/ws/portalfiles/portal/5149389/Hophmayer_2006_Wastewater+Management+Strategy+centralized+v.+decentralized+technologies+for+small+communities.pdf

ISO 14040:2006. Environmental Management—Life Cycle Assessment—Principles and Framework. Available online: https://www.iso.org/standard/37456.html (accessed October 28, 2022).

Iribarnegaray, M. A., M. S. Rodriguez Alvarez, D. Oliva, W. Tejerina, V. I. Liberal, and L Seghezzo. 2016. Tratamiento descentralizado de aguas residuales doméstivas en el área metropolitana del Valle de Lerma, provincia de Salta, Argentina. *Energías Renovables y Medio Ambiente* 37:35–41. Available online: http://eprints.natura.unsa.edu.ar/id/eprint/822

Iribarnegaray, M.A., M.S. Rodriguez-Alvarez, L.B. Moraña, W.A. Tejerina, and L Seghezzo. 2018. Management challenges for a more decentralized treatment and reuse of domestic wastewater in metropolitan areas. *Journal of Water, Sanitation & Hygiene for Development* 8(1):113–122. https://doi.org/10.2166/washdev.2017.092.

Jackson, H. B. 1996. Global Needs and Developments in Urban Sanitation. In *Low-Cost Sewerage*, ed. D. Mara. Chichester: John Wiley & Sons. Available online: https://sleigh-munoz.co.uk/wash/Sewerage/articles/WHO_EOS_98.5_pp249-262.pdf

Jenkins, M. W., S. K. Tiwari, and J. Darby. 2011. Bacterial, viral and turbidity removal by intermittent slow sand filtration for household use in developing countries: Experimental investigation and modeling. *Water Research* 45:6227–6239. https://doi.org/10.1016/j.watres.2011.09.022

Jia, S., and X. Zhang. 2020. Biological HRPs in Wastewater. In *High-Risk Pollutants in Wastewater*, ed. H. Ren and X. Zhang, 41–78. Elsevier. https://doi.org/10.1016/B978-0-12-816448-8.00003-4.

Jones, D.S., A.Q. Armstrong, M.D. Muhlheim, and B.V. Sorensen. 2000. Integrated Risk Assessment/Risk Management as Applied to Decentralized Wastewater Treatment: A High-Level Framework. Environmental Sciences Division Publication No. 5032. U.S. Department of Energy. Available online: https://technicalreports.ornl.gov/cpr/misc/108367_.pdf (accessed October 10, 2022).

Kadir, K., and K. L. Nelson. 2014. Sunlight mediated inactivation mechanisms of *Enterococcus faecalis* and *Escherichia coli* in clear water versus waste stabilization pond water. *Water Research* 50:307–317. https://doi.org/10.1016/j.watres.2013.10.046

Lane, J. L., D. W. de Haas, and P. A Lant. 2015. The diverse environmental burden of city-scale urban water systems. *Water Research* 81:398–415. https://doi.org/10.1016/j.watres.2015.03.005

Lautze, J., E. Stander, P. Drechsel, A. K. da Silva, and B. Keraita. 2014. Global Experiences in Water Reuse. In *CGIAR Research Program on Water, Land and Ecosystems (WLE)*. Colombo: International Water Management Institute (IWMI). Available online: https://www.pseau.org/outils/ouvrages/cgiar_iwmi_global_experiences_in_water_reuse_2014.pdf

Lens, P., G. Zeeman, and G. Lettinga. 2005. *DecDecentralizednitation and Reuse: Concepts, Systems and Implementation*. London: IWA Publishing. https://doi.org/10.2166/9781780402949

Lourenço, N., and L.M Nunes. 2020. Review of dry and wet decentralized sanitation technologies for rural areas: Applicability, challenges and opportunities. *Environmental Management* 65:642–664. https://doi.org/10.1007/s00267-020-01268-7

Malila, R., S. Lehtoranta, and E. L. Viskari. 2019. The role of source separation in nutrient recovery comparison of alternative wastewater treatment systems. *Journal of Cleaner Production* 219:350–358. https://doi.org/10.1016/j.jclepro.2019.02.024

Massoud, M. A., A. Tarhini, and J. A Nasr. 2009. Decentralized approaches to wastewater treatment and management: Applicability in developing countries. *Journal of Environmental Management* 90(1):652–659. https://doi.org/10.1016/j.jenvman.2008.07.001

Morató, J., F. Codony, O. Sánchez, L. M. Pérez, J. García, and J. Mas. 2014. Key design factors affecting microbial community composition and pathogenic organism removal in horizontal subsurface flow constructed wetlands. *Science of the Total Environment* 481:81–89. https://doi.org/10.1016/j.scitotenv.2014.01.068

Nguyen, M. T., J. T. Jasper, A. B. Boehm, and K. L. Nelson. 2015. Sunlight inactivation of fecal indicator bacteria in open-water unit process treatment wetlands: Modeling endogenous and exogenous inactivation rates. *Water Research* 83:282–292. https://doi.org/10.1016/j.watres.2015.06.043

Novotny, V., and P. Brown. 2007. *Cities of the Future: Towards Integrated Sustainable Water and Landscape Management.* London: IWA Publishing. https://doi.org/10.2166/9781780405308

Paraskevas, P.A., D.L. Giokas, and T.D. Lekkas, 2002. Wastewater management in coastal urban areas: The case of Greece. *Water Science and Technology* 46(8):177–186. Available online: https://pubmed.ncbi.nlm.nih.gov/12420981/

Pundsack, J., R. Axler, R. Hicks, J. Henneck, D. Nordman, and B. McCarthy. 2001. Seasonal pathogen removal by alternative on-site wastewater treatment systems. *Water Environment Research* 73:204–212. https://doi.org/10.2175/106143001x139182

Redder, A., M. Dürr, G. Daeschlein, O. Baeder-Bederski, C. Koch, R. Müller, M. Exner, and M Borneff-Lipp. 2010. Constructed wetlands—Are they safe in reducing protozoan parasites? *International Journal of Hygiene and Environmental Health* 213(1):72–77. https://doi.org/10.1016/j.ijheh.2009.12.001

Reed, R. A. 1996. Selecting Communities for Sewerage. In *Low-Cost Sewerage*, ed. D. Mara. Chichester: John Wiley & Sons. Available online: https://www.researchgate.net/publication/230888006_Low-Cost_Sewerage

Reynaud, N., and C Buckley. 2015. Field-data on parameters relevant for design, operation and monitoring of communal decentralized wastewater treatment systems (DEWATS). *Water Practice and Technology* 10(4):787–798. https://doi.org/10.2166/wpt.2015.097

Roefs, I., B. Meulman, J. Vreeburg, and M. Spiller, M. 2017. Centralised, decentralised or hybrid sanitation systems? Economic evaluation under urban development uncertainty and phased expansion. *Water Res* 109:274–286. https://doi.org/10.1016/j.watres.2016.11.051

Rodríguez, L. G. 2009. El tratamiento descentralizado de aguas residuales doméśticas como alternativa sostenible para el saneamiento periurbano en Cuba. *Ingeniería hidráulica y ambiental* 30(1):29–35. Available online: https://riha.cujae.edu.cu/index.php/riha/article/view/87

Romeiko, X. X. 2020. Assessing health impacts of conventional centralized and emerging resource recovery-oriented decentralized water systems. *International Journal of Environmental Research and Public Health* 17:973. https://doi.org/10.3390/ijerph17030973

Sasidharan, S., S. Torkzaban, S.A. Bradford, R. Kookana, D. Page, and P.G. Cook. 2016. Transport and retention of bacteria and viruses in biochar-amended sand. *Science of the Total Environment* 548:100–109. https://doi.org/10.1016/j.scitotenv.2015.12.126

Schoen, M. E., X. Xue, A. Wood, T. R. Hawkins, J. Garland, and N. J. Ashbolt. 2017. Cost, energy, global warming, eutrophication and local human health impacts of community water and sanitation service options. *Water Research* 109:186–195. https://doi.org/10.1016/j.watres.2016.11.044

Shehabi, A., J. R. Stokes, and A Horvath. 2012. Energy and air emission implications of a decentralized wastewater system. *Environmental Research Letters* 7(2):1–6. Available online: https://iopscience.iop.org/article/10.1088/1748-9326/7/2/024007/pdf

Sheikh, M., I. Fakhrul, and K. Zahurul. 2019. World's Demand for Food and Water: the Consequences of Climate Change. In *Desalination - Challenges and Opportunities*, ed. Farahani, Mohammad Hossein Davood Abadi, Vatanpour, Vahid, Taheri, Amir Hooshang. IntechOpen. https://doi.org/10.5772/intechopen.85919.

Shingare, R.P., P.R. Thawale, K. Raghunathan, A. Mishra, and S. Kumar. 2019. Constructed wetland for wastewater reuse: Role and efficiency in removing enteric pathogens. *Journal of Environmental Management* 246:444–461. https://doi.org/10.1016/j.jenvman.2019.05.157

Silverman, A. I., B. M. Peterson, A. B. Boehm, K. McNeill, and K. L. Nelson. 2013. Sunlight inactivation of human viruses and bacteriophages in coastal waters containing natural photosensitizers. *Environmental Science & Technology* 47:1870–1878. https://doi.org/10.1021/es3036913

Singh, A., M. Sawant, S. J. Kamble, M. Herlekar, M. Starkl, E. Aymerich, and A Kazmi. (2019). Performance evaluation of a decentralized wastewater treatment system in India. *Environmental Science and Pollution Research* 26(21):21172–21188. https://doi.org/10.1007/s11356-019-05444-z

Sjogren, R.E. 1994. Prolonged survival of an environmental *Escherchia coli* in laboratory soil microcosms. *Water, Air, and Soil Pollution* 75:389–403. https://doi.org/10.1007/BF00482948

Starkl, M., N. Brunner, S. Das, and A. Singh. 2022. Sustainability assessment for wastewater treatment systems in developing countries. *Water*, 14(2):241. https://doi.org/10.3390/w14020241.

Stevik, T.K., K. Aa, G. Ausland, and J. F. Hanssen. 2004. Retention and removal of pathogenic bacteria in wastewater percolating through porous media: A review. *Water Research* 38:1355–1367. https://doi.org/10.1016/j.watres.2003.12.024

Sundaravadivel, M., and S Vigneswaran. (2001). Wastewater collection and treatment technologies for semi-urban areas of India: A case study. *Water Science and Technology* 43(11): 329–336. https://doi.org/10.2166/wst.2001.0699

Suneethi, S., G. Keerthiga, R. Soundhar, M. Kanmani, T. Boobalan, D. Krithika, and L Philip. 2015. Qualitative evaluation of small scale municipal wastewater treatment plants (WWTPs) in South India. *Water Practice and Technology* 10(4): 711–719. https://doi.org/10.2166/wpt.2015.087

Tchobanoglous, G., L. Ruppe, H. Leverenz, and J Darby. 2004. Decentralized wastewater management: Challenges and opportunities for the twenty-first century. *Water Science and Technology: Water Supply* 4(1): 95–102. https://doi.org/10.2166/ws.2004.0011

UNEP GLOBAL PROGRAMME OF ACTION (GPA). 2000. Environmentally Sound Technologies in Wastewater Treatment. Available online: http://www.unep.or.jp/ietc/Publications/Freshwater/SB_summary/index.asp (accessed October 28, 2022).

Valencia, E., I. J. Silva, and C. P Narváez. 2010. Sistemas descentralizados integrados y sostenibles para el tratamiento de aguas residuales domésticas. *Ingeniería y Región* 7: 65–72. https://doi.org/10.25054/22161325.801

Vandenhove, H., R. Merckx, H. Wilmots, and K. Vlassak. 1991. Survival of *Pseudomonas fluorescens* inocula of different physiological stages in soil. *Soil Biology and Biochemistry* 23:1133–1142. https://doi.org/10.1016/0038-0717(91)90025-F

Wagner, M, and I Strube. 2005. Risk management in waste water treatment. *Water Science and Technology* 52(12):53–61. Available online: https://iwaponline.com/wst/article-abstract/52/12/53/12458/Risk-management-in-waste-water-treatment?redirectedFrom=PDF

Wang, M., J. Zhu, and X. Mao. 2021. Removal of pathogens in on-site wastewater treatment systems: A review of design considerations and influencing factors. *Water* 13:1190. https://doi.org/10.3390/w13091190

Wilderer, P. A., and D Schreff. 2000. Decentralized and centralized wastewater management: A challenge for technology developers. *Water Science and Technology* 41(1): 1–8. https://doi.org/10.2166/wst.2000.0001

Winfield, M. D., and E. A. Groisman. 2003. Role of non-host environments in the lifestyles of *Salmonella* and *Escherichia coli*. *Applied and Environmental Microbiology* 69:3687–3694. https://doi.org/10.1128/AEM.69.7.3687-3694.2003

World Health Organization (WHO) and UNICEF. 2015. *Progress on Sanitation and Drinking Water: 2015 Update and MDG Assessment.* Geneva: World Health Organization. Available online: https://www.who.int/publications/i/item/9789241509145

World Health Organization (WHO). 2017. *Progress on Drinking Water, Sanitation and Hygiene: 2017 Update and SDG Baselines.* Geneva: World Health Organization. Available online: https://policycommons.net/artifacts/421754/progress-on-drinking-water-sanitation-and-hygiene/1392737/

World Health Organization (WHO). 2021. Progress on household drinking water, sanitation and hygiene, 2000–2020: five years into the SDGs. Available online: https://www.who.int/publications/i/item/9789240030848 (accessed May 4, 2022).

World Health Organization (WHO). 2022a. WHO publishes list of bacteria for which new antibiotics are urgently needed. Available online: https://www.who.int/news/item/27-02-2017-who-publishes-list-of-bacteria-for-which-new-antibiotics-are-urgently-needed (accessed October 10, 2022).

World Health Organization (WHO). 2022b. *World Health Statistics 2022: Monitoring Health for the SDGs, Sustainable Development Goals.* Geneva: World Health Organization; 2022. Licence: CC BY-NC-SA 3.0 IGO. Available online: https://apps.who.int/iris/bitstream/handle/10665/356584/9789240051140-eng.pdf

World Water Assessment Programme. 2020. *The United Nations World Water Development Report 2020: Water and Climate Change.* Paris: United Nations Educational Scientific and Cultural Organization (UNESCO). Available online: https://www.unwater.org/publications/un-world-water-development-report-2020

Wu, S., P. N. Carvalho, J. A. Müller, V. R. Manoj, and R. Dong. 2016. Sanitation in constructed wetlands: A review on the removal of human pathogens and fecal indicators. *Science of the Total Environment* 541:8–22. https://doi.org/10.1016/j.scitotenv.2015.09.047

Xue, X., S. Cashman, A. Gaglione, J. Mosley, L. Weiss, X. Ma, J. Cashdollar, and J. Garland. 2019. Holistic analysis of urban water systems in the greater Cincinnati region: (1) life cycle assessment and cost implications. *Water Research* 2:100015. https://doi.org/10.1016/j.wroa.2018.100015

Zaleski, K. J., K. L. Josephson, C. P. Gerba, and I. A. N. L Pepper. 2005. Survival, growth, and regrowth of enteric indicator and pathogenic bacteria in biosolids, compost, soil, and land applied biosolids. *Journal of Residuals Science and Technology* 2:49–63.

Zhao, J.L., B. Li, P. Lv, J. Hou, Y. Qiu, and X. Huang, X. 2022. Distribution of antibiotic resistance genes and their association with bacteria and viruses in decentralized sewage treatment facilities. *Frontiers of Environmental Science & Engineering* 16(3):35. https://doi.org/10.1007/s11783-021-1469-4.

7 Social Aspects and Public Acceptance of Decentralized Treatment

Chinthala Sumanth[a], *Bhukya Gopal*[b], *Manojkumar Y.*[b], *and Sridhar Pilli*[a]

[a] National Institute of Technology (NIT), Warangal, India

[b] Water and Environment Division, Department of Civil Engineering, NIT Telangana, India

7.1 INTRODUCTION

Decentralized treatment simply refers to placing a scaled water or wastewater treatment facility close to a supply, a requirement, or ideally both. Decentralization has long been a desirable treatment approach, but its adoption has been constrained by inadequacies in engineering and technology. Remote villages have frequently been left without service, because determining the cost of building pipelines to cover distant locations is one of the most expensive components of conventional, centralized treatment. The cost of providing such regions with service has decreased because of the placement of treatment plants precisely where they are required. In addition, economies of scale are no longer necessary to achieve cost-efficiency because of recent developments in compact wastewater treatment and desalination.

The circular economy is supported by decentralized technology, which entails treating wastewater locally (in homes, neighbourhoods, or small towns), encouraging the local recovery of water and nutrients for new purposes, and challenging the conventional method of disposing of wastewater away from sources.

Decentralized plants might nevertheless incur installation, maintenance, and location-based expenses despite their benefits (Gómez-Román et al., 2021).

Decentralized wastewater treatment options include natural treatment methods, aerobic and anaerobic systems, as well as combination (natural/aerobic/anaerobic) systems (Gómez-Román et al., 2020).

7.2 TYPES OF DECENTRALIZATION

Decentralization is classified according to the population at different scales:

1. Suburb/district level. The system might also be referred to as centralized when applied to small towns or rural regions at the level of a satellite township, district or suburb. The entire system may be referred to as a "hybrid" system if it is only implemented in a few districts or suburbs in medium- or large-sized population centres with the current centralized system.
2. Neighbourhood level. It can include gated neighbourhoods, tiny districts, and regions that are supplied by gravity or vacuum sewers in this group.
3. On-site level. Within the limits of each building's property.

DOI: 10.1201/9781003376415-9

7.2.1 Concern for Ecosystem and Human Health

The water deficit is an environmental issue that could be potentially solved by decentralized wastewater treatment plants. There must be some level of public understanding or care about environmental concerns for that technology to be accepted; however, this is not sufficient. Decentralized wastewater treatment facilities enable water and nutrient recovery and reuse, boosting the circular economy (Gómez-Román et al., 2020).

Decentralized wastewater treatment systems are a practical and essential alternative to centralized systems for managing wastewater, helping to reduce negative environmental effects, facilitating resource recovery, and giving residents in rural and peri-urban areas access to basic sanitation (Gómez-Román et al., 2021).

Due to the reduced wastewater flow treated by decentralized systems, which facilitate pollutant reduction and reduce the environmental impact of pollutants on variables like dissolved oxygen, chemical oxygen demand, and biochemical oxygen demand, there is also the environmental benefit of reduction in the amount of pollution that water bodies receive after treatment (Bernal et al., 2021). Further, the public is not informed about the technologies so that they can have an opinion (Valentin and Bogus, 2015).

7.2.2 The Public's Current Perception

Despite the potential benefits of these systems, their effective deployment depends on societal acceptability. Public individuals consider both the positive and negative features of these systems, including the economic implications. Regardless of the findings about the advantages and disadvantages of decentralized systems, in order for the public to accept these systems, people must believe that some aspect of the present water management system needs to be improved.

Strategies to address the issue of water stress will likely receive little or no support if the general public does not consider environmental concerns to be a priority. When presenting information to the public, it is important to keep in mind that several factors may affect the impact that information may have (to a greater or lesser extent). Presenting positive messages while simultaneously addressing certain drawbacks or less favourable aspects increases trust, leading people to be more accepting of decentralized treatment systems (Gómez-Román et al. 2020).

Large-scale water and wastewater infrastructure is planned and built mostly by local, regional, and state governments. Management techniques need to be interdisciplinary if water treatment systems are to be sustained. Nowadays, many people believe that the future of urban water management lies in the adoption of decentralized management systems. These facilities support the circular economy by permitting the recovery of water for new purposes and allowing for the differential treatment of wastewater that has been isolated from its source (Gómez-Román et al. 2020).

7.3 SOCIAL AND SOCIO-ECONOMIC IMPLICATIONS

For a centralized system, expenditures account for around 60% of the entire budget when the three fundamental parts of a wastewater management system – collection, treatment, and disposal – are considered. This component is scaled back and focused mostly on wastewater treatment and disposal in decentralized systems.

The goal of the decentralization strategy is to create systems that are more environmentally friendly, socially responsible, and financially stable than centralized traditional systems (Gómez-Román et al. 2021).

Completely decentralized systems are uncommon, particularly in areas with modern housing developments and strong infrastructure. However, it is a common occurrence in rural parts of underdeveloped nations, where there are few basic sanitation services. The use of a well-constructed and maintained pit latrine or septic tank facility for each family might be regarded as a good sanitation practice under such circumstances. It serves as an illustration of a totally decentralized system. Decentralized wastewater systems might be a practical substitute for the traditional centralized system in places with developed infrastructure, particularly when it comes to enhancing or adapting existing systems. "Hybrid" systems can take on a wide variety of configurations and modifications. Decentralized solutions that are both technically and aesthetically sound and acceptable are now possible thanks to the development of new treatment technologies. Decentralized applications are essential for new residential or commercial complexes where the infrastructure is still being built or won't be finished for some time. Roads, water supplies, and particularly wastewater and drainage systems are developed in many nations and regions years after the construction of homes. Typically, this is brought on by a lack of resources and subpar planning techniques. Decentralized applications are necessary for these situations to protect the environment and general health.

7.4 POLICIES APPLICABLE TO DECENTRALIZED SYSTEMS

A wastewater treatment system's level of centralization or decentralization will mostly rely on how industrialized the region is where it is built. In order to ensure the longevity of these networks, highly industrialized regions typically develop centralized systems linking many areas and communities. This is because they have the necessary financial and technological resources. Decentralization, on the other hand, is viewed as a workable solution for poor basic sanitation coverage difficulties in developing nations, which typically lack full coverage or the necessary governmental, economic, and technical support required to ensure system longevity (Bernal et al., 2021).

The issue with traditional civil engineering projects is that they were never able to combine the management of drinking water, energy, and wastewater (Chirisa et al., 2017). The National Urban Sanitation Policy (NUSP) was established by the government to encourage the use of low energy-intensive decentralized technology for wastewater treatment in order to address these restrictions in nations like India (Jung et al., 2018). General studies were covered in some of the assessments, while many others explored policy and application concerns for decentralized systems in developing nations. For impoverished countries like Zimbabwe, Chirisa et al. (2017) claim that the largest barrier to implementing decentralized systems is that existing government policies and legislation are built on centralized infrastructure. As a result, there aren't any institutional or regulatory frameworks that support or encourage the use of decentralized systems.

Therefore, developing nations need to have appropriate rules and institutional frameworks that incorporate sanitation policies consistent with decentralized technologies to a larger extent for the effective implementation or acceptance of decentralized wastewater systems (Nansubuga et al., 2016). Decentralized systems have been successfully modelled as acceptable technology for wastewater treatment in two countries: India and Brazil. The National Basic Sanitation Plan (Plansab), which was introduced in 2013, encouraged private investment in sanitation, allowing additional actors to enter the fray who were in favour of the development of decentralized systems, according to the assessment by Ferreira et al. (2021). According to Brunner et al. (2018), the National Urban Sanitization Policy of 2008 in India paved the path for the use of decentralized technology. Other programmes, like the Integrated Urban Sanitation Programs, introduced in 2009, and the Clean India Mission, introduced in 2014, enhanced this approach even more. As a result, Indian towns used these legal instruments to develop sanitation plans based on a legally defined general framework for planning, implementing, and managing decentralized wastewater systems.

The United Nations Economic and Social Commission for Asia and the Pacific (UNESCAP, 2015) stated the following principles and various policy frameworks.

- Vision and Setup Targets
- Assess Sanitation Strategies and Socio-Economic Impacts
- Analyse Institutions and Partners
- Analyse Costs and Benefits

- Assess the Strategic Future
- Choose the Technology System
- Increase the Demand
- Ensure Financial and Investment Climate from Government, Donors, and Entrepreneurs
- Ensure Exit Strategy and Sustainability
- Evaluate Implementation

However, there are still gaps in the attempts of developing countries to embrace decentralized treatment systems, and one thing that must be fixed in most countries is policy. Government policies have a significant impact on how a nation manages its wastewater, therefore broad policy reforms can aid in the acceptance and implementation of decentralized systems for long-term wastewater management in developing nations. Future studies might thus focus on factors that facilitate the successful adoption of the systems rather than just legislation (Muzioreva et al., 2022).

7.5 ROLE OF STAKEHOLDERS AND ACTORS IN THE PROCESS

Programs for the development of sustainable infrastructure and sanitation must coordinate and streamline a wide range of stakeholders and resources. Various stakeholders should actively participate throughout the whole development process, from planning to implementation to monitoring to final evaluation. Participation enhances a project's performance and sustainability. Ownership ensures commitment and engagement from stakeholders, which lowers the cost of oversight.

The systematic inclusion of many stakeholder groups is necessary for efficient, economical, and sustainable implementation. Stakeholder groups include the following:

- Primary stakeholders are the residents and immediate consumers of the adopted interventions.
- Secondary stakeholders, or organizations, have a direct or indirect role in the programme. Leading organizations (public, NGOs, etc.), planning authorities, and departments of health and the environment are among them.
- Tertiary stakeholders are companies that offer specialized services for building, upkeep, and sludge management.

7.6 CASE STUDIES

7.6.1 Decentralized Wastewater Systems in Bengaluru, India

Bengaluru is one of the largest cities in India. The water source for this city is from Cauvery River, which is 100 km away and 300 m downhill. Bengaluru is also facing water pollution due to wastewater discharge into lakes. In 2004, due to an increase in water crisis and pollution, KSPCB (Karnataka State Pollution Control Board) issued a ZLD (Zero Liquid Discharge) order. According to this, an apartment complex with more than 50 residential units or a built-up area of more than 5,000 m^2 should install on-site STP (Sewage Treatment Plant) and reuse 100% treated water. Due to the ZLD order, around 2,200 STPs were reported in the database (Kuttuva et al., 2018).

According to Kuttuva et al. (2018), the STPs are characterized by the level of compliance and data collected from field visits. This study reported the limitations of the ZLD model in Bengaluru. It was observed that the design and technology of STP are optional and in the hands of the builder of the apartment. Most of the STPs are under capacity and the residents of the building were not involved in the design. Regular monitoring of treatment and reuse was not noticed. In some cases, the cost of the treatment is more than the availability of freshwater due to borewells and other sources. Continuous treatment and reuse of water are observed in a few STPs where tanker water is used, which is expensive for gardening and flushing purpose. High treatment costs made several

STPs not run 24×7. Generally, treated water is used for sprinkling and flushing purposes. Most of the apartments have less than 25% free space and there are three months of rainfall which leads to the break in 24×7 and reuse of treated water. Due to less space and other factors, 100% reuse is not feasible.

The authorities could have permitted to discharge of treated water in waterbodies instead of ZLD. There is a lack of awareness about environmental pollution among the people leading the regular operation and maintenance issues in STPs. It was observed that, due to the limitations in the system, decentralized wastewater treatment in Bengaluru is identified as not effective in implementation and monitoring (Kuttuva et al., 2018).

7.6.2 Decentralized Wastewater System in Harare, Zimbabwe

Planners, engineers, and systems thinkers are compelled to re-evaluate what may function as a sustainable wastewater system in Zimbabwe, with specific reference to the ever-expanding Harare metropolitan region, due to the system's present limitations and the country's massive urbanization. A decentralized system for collecting and treating home wastewater is the recommended framework of operation. Decentralized wastewater management systems in Zimbabwe are mostly in rural regions where pit latrines are often utilized. It is normal practice in areas like Gweru, Redcliff, Nemanwa, and Mupandawana to handle wastewater by decentralized treatment.

The decentralized wastewater system in Gweru began operation in 1994. The reuse of wastewater for agriculture in the municipality of Redcliff is proof of a decentralized wastewater treatment and management system. A typical wet technology utilized in places with big plot sizes is septic tanks. Aquaculture, man-made wetlands, and duckweed pond systems are all employed at Redcliff. The utilization and accessibility of duckweed pond systems in Mupandawana and Nemanwa offer a chance for the establishment of decentralized wastewater treatment and management systems.

The administrative, commercial, and communications capital of Zimbabwe is the city of Harare, which is located at a height of 1,483 metres. The city has a population of 2,098,199. Currently, Harare's wastewater management and treatment systems are centralized. With the help of flush toilets, Harare's sewage system provides services to almost 1.8 million residents in the city's official settlements. However, this system is now obsolete and cannot be used in informal settlements, where open defecation and pit latrines are common.

The sewage is sent to the overburdened Crowborough and Firle's activated sludge facilities. These facilities are 44% overloaded since their entire design capacity is 208,000 m^3/day while their total daily current inflows are closer to 300,000 m^3/day. The primary water supplies for the city, Lakes Chivero and Manyame, are reached by large amounts of wastewater that has not been properly cleaned and is being dumped into the Marimba and Mukuvisi Rivers.

First preference was placed on an on-site treatment and reuse on the assumption of a differentiated and integrated strategy. The second choice was to handle everything that couldn't be handled on-site at a decentralized level. The third option was to handle the rest on a centralized level. Regarding the decentralized wastewater management idea, Harare may implement several different tactics. As decentralized wastewater management systems, septic tanks, pit latrines, and urine diversion are some of the solutions that are becoming more and more common. Less-concentrated wastewater can also be treated naturally using ponds with algae and duckweed as well as artificial wetlands that also gather protein-rich biomass. Other methods for stabilizing organic materials include the use of maturation ponds and anaerobic pre-treatment. In Harare, decentralized systems also incorporate algae-based technologies.

Rethinking the existing approach to urban water management is prompted by the socio-economic and environmental issues that now affect Harare. Decentralization must be included as a practical alternative for managing wastewater in metropolitan areas, and reuse activities must be regulated by establishing quality standards, among other things, in order to connect the theoretical framework with actual actions. The social, financial, and environmental benefits of decentralization become crucial considerations when considering this type of scheme in developing countries' urban water management plans, such as for Harare, particularly in the peri-urban areas, informal settlements, and southern suburbs where wastewater collection and/or treatment is not available. This conclusion is based on a review of the state of the art and experiences with decentralization. Given the urban form of Harare, it may be possible to implement decentralization through a semi-centralized plan, as there aren't many elements in the planning policies and regulatory framework that would allow for management other than the traditional "end-of-the-pipe" solutions and the use of standard technologies in centralized systems (Chirisa et al., 2017).

7.6.3 Centralized versus Decentralized Wastewater Treatment Plants in Bangkok, Thailand

Eight centralized wastewater treatment plants (WWTPs; Bangsue, Chatuchak, Chongnonsi, Dindaeng, Nongkaem, Rattanakosin, Sipraya, and Thungkru), 12 decentralized WWTPs (Bangbua, Bangna, Bonkai, Huai-khwang, Hua-mark, KhlongChan, KhlongToei, RamIntra, RomKlao, ThaSai, and Tungsonghong I and II) are now present in Bangkok, the capital of Thailand. Dewatering is used to remediate the sludge in the decentralized WWTPs for dry organic waste. Sludge is delivered by truck to Nongkeam WWTP for centralized wastewater treatment after thickening, where it is converted into biogas and fertilizer. Sludge is sun-dried for decentralized wastewater treatment and then utilized as landfill. Therefore, composting should be used to remediate sludge.

Construction of a centralized WWTP takes two years and costs 3,358.27 million THB; a decentralized WWTP takes one year and costs 118.95 million THB, according to the Department of Drainage and Sewerage of the Japan International Cooperation Agency. Compared to decentralized treatment, the average input of centralized treatment is greater and includes transportation, water supply, and energy. Due to the centralized WWTPs' greater treatment efficacy, the average output of the centralized treatment, including air (CO_2, CH_4) and water (BOD, total P) emissions, is lower. The centralized treatment produces less-direct greenhouse gas (GHG) emissions than the dispersed treatment. However, the centralized treatment produces more indirect GHG emissions because of its increased power usage. The centralized treatment generates more sludge than the decentralized treatment, which accounts for the greater levels of heavy metals in those cases. The centralized treatment scenarios have higher overall costs than the decentralized treatment scenarios (Prateep Na Talang et al., 2022).

7.6.4 Public-Private Partnerships for Wastewater Treatment in Rural Areas: Changshu, China

Rural development in the People's Republic of China (PRC) has improved and the income level of the residents has also increased. Although the quality of life has improved in rural areas, environmental rehabilitation still needs to be addressed. In 2017, China introduced the "rural toilet revolution" and an action plan to improve rural sanitation. Domestic wastewater treatment is the main aim of rural sanitation. Wastewater generated from toilets, kitchens, and washing activities causes environmental pollution. Collection and treatment of wastewater at centralized facilities from different villages requires a large network of pipelines and is not economical. In order to address the issue, decentralized wastewater treatment is an appropriate option which is more suitable for rural areas. It requires a small network of pipelines, and on-site treatment is an economical process. The challenges of decentralized wastewater treatment are treated effluent quality and monitoring of operation and maintenance of treatment units.

Changshu is a county-level city under the administration of Suzhou city. Changshu is facing water pollution due to wastewater discharges, making it a case study for developing rural wastewater management. The city invested in 180 sets of decentralized wastewater facilities. At this stage, the city identified a few problems in the system; the percentage of decentralized facilities was less than 40% in good condition and the operation rate was less than 15%. In order to improve sanitation, the Changshu treasury provided the operating cost of the decentralized wastewater treatment by eliminating the financial barriers. The authorities collected wastewater treatment fees from the residents based on their water consumption and used them in the construction of decentralized wastewater treatment facilities to improve the sanitation of rural areas. As a result, the city's decentralized treatment plants increased to 95% and the operating rate to 92%. The problem identified in this case is frequent repair and maintenance, which increased high operation costs in decentralized systems.

In order to overcome the problem, the city raised the construction cost of the households; but the situation is not improved. To improve the quality of sanitation, the Research Centre for Eco-Environment Sciences of the Chinese Academy of Sciences (RCEES) worked with the Japanese household Johkasou in Changshu city. A demo project implemented a total of 54 sets of Johkasou and collected wastewater from 102 households. This project reduced the requirement of pipes, construction, installation, and operation costs of the project. The Japanese household Johkasou project brings decentralized wastewater treatment to an advanced level by reducing construction costs. Most of the facilities met the effluent standards. Based on the satisfactory results from Japanese Johkasou technology, Changshu and RCEES developed a plan to operate decentralized wastewater treatment through a public-private partnership (PPP).

This plan increased the PRC's manufacturing technologies for decentralized facilities. It also provided a system for installation, operation, and third-party monitoring in decentralized wastewater treatment. Changshu implemented a three-year action plan, in which phase 1 of the PPP project, involving 330 villages, planned to collect 4,192.4 tons of sewage per day. With the success of phase 1, Changshu implemented comprehensive wastewater management in villages and towns. Regular monitoring of operations and effluent standards minimizes system failures and results in effective wastewater management. The PPP models should focus further on the quality and efficiency of the project to improve sanitation in rural areas (ADBI, 2022).

7.7 STRATEGIES AND APPROACHES FOR PUBLIC ACCEPTANCE

If the public accepts a decentralized system, they should be made aware of how the decentralized system works. The approaches include the following:

1. Clarifying the responsibilities assigned to different government levels.
2. Ensuring all responsibilities are sufficiently funded.
3. Strengthening subnational fiscal autonomy to enhance accountability at every level.
4. Supporting subnational capacity building.
5. Building adequate coordination mechanisms among every level of government.
6. Supporting cross-jurisdictional co-operation.
7. Strengthening innovative and experimental governance, and promoting citizen engagement.
8. Allowing for and making the most of asymmetric decentralization-related arrangements.
9. Consistently improving transparency, enhancing data collection, and strengthening performance monitoring.
10. Strengthening national regional development policies and equalization of the systems.

Additionally, legal frameworks must be explained to the public about the jurisdictions under which the conflicts are resolved. For instance, if a decentralized treatment plant has odour problems, then the residents should know the course of action. Additionally, the residents should also be made aware of who will pay for the maintenance and other operational costs associated with the decentralized unit.

7.8 BENEFITS OF PUBLIC ACCEPTANCE OF DECENTRALIZED SYSTEMS

7.8.1 Sharing Benefits

The whole intention of encouraging decentralized systems is to ensure that the overall efficiency is increased and the system is made greener. However, in most cases, the economic benefits will go to the city administration. If those benefits are shared with the public, then public perception may change. For instance, people did not prefer buying electric vehicles, until the subsidy on electric vehicles motivated many people to accept the technology. Similarly, if benefits like reducing the property tax by 5%, for example, are introduced, the public may accept decentralized treatment unit.

7.8.2 Providing Low-Cost Alternatives

Many households use one pipe system where the same water is used for flushing and other domestic activities. If the decentralized system can connect the flush tanks with a pipe system and make people reuse the water for flushing, a significant amount of maintenance costs shall be reduced for the user. Offering such benefits will help to convince more and more people of the decentralized systems.

7.8.3 Providing Information

Since the public may not have a complete idea about the degree of environmental problems, they may not accept them. Even if they have an idea about the environmental problems, if the solutions are not properly explained to them, those solutions may be rejected. For instance, if the reused water from a decentralized technology is made available for domestic use, questions may arise about the hygiene and safety of the residents if the reused water is used in their homes. If proper guidelines and protocols along with the monitoring data are not presented, then the common public may not accept the reused water for their domestic purposes.

7.8.4 Quantification

Generally, social acceptance depends on the "majority". Many people question how many people have already accepted this technology and may ask for the data for the same. And if they deny the technology, sociology experts have to investigate why they are denying it. If the reasons for denial are quantified along with other factors, the scientific community may find suitable alternatives for ensuring social acceptance (Wilson et al., 1989).

7.8.5 Hybrid Systems

When the public knows about various technologies, public acceptance may lead to a debate where they start asking about why one particular technology is preferred over another one. Under such cases, economic evaluation and the presentation of the results are mandatory. People may ask for hybrid systems which may be more suitable, but the municipal authorities always have restrictions and they may have to choose a technology based on many other related factors. Hence, economic evaluation is also needed to convince people of the decentralized systems (Roefs et al., 2017). The methodology for ensuring social acceptance of decentralized technologies is shown in Figure 7.1. The double arrows indicate the parallel activities which are to be carried out simultaneously to ensure that people accept the technology.

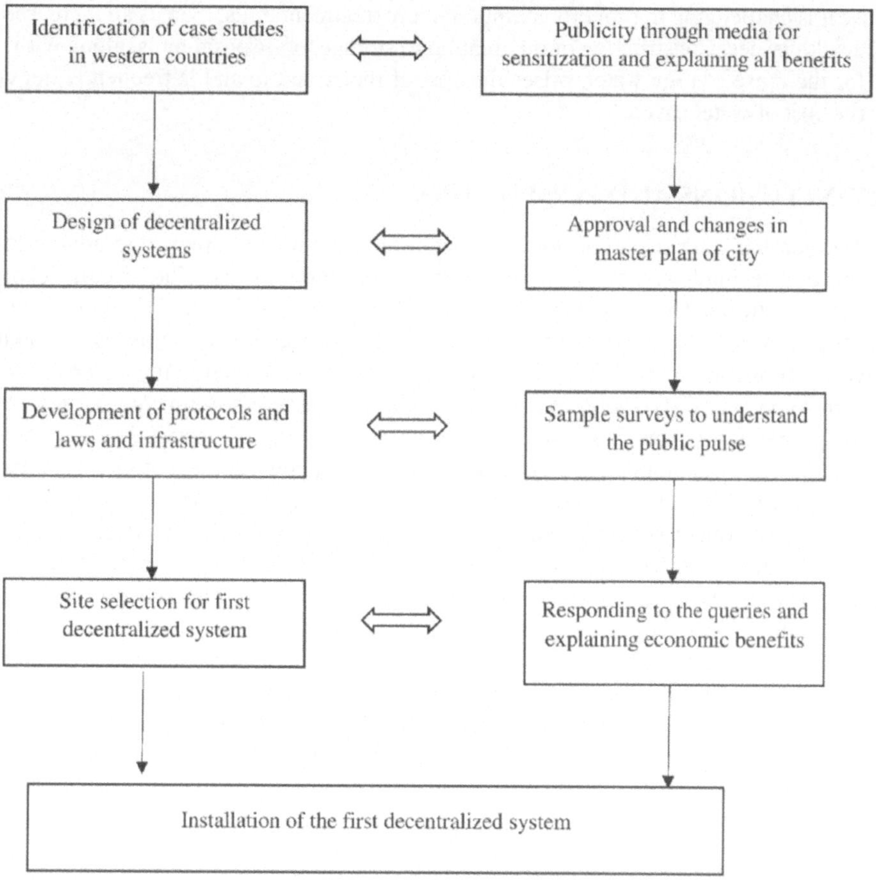

FIGURE 7.1 Social acceptance of decentralized systems: The methodology.

7.9 CHALLENGES OF PUBLIC ACCEPTANCE IN DECENTRALIZED SYSTEMS

Social acceptance of decentralized systems among the public is generally dependent on many factors, for example, perception, property value, sharing benefits, providing low-cost alternatives, providing information, quantification, and hybrid systems. The perception of the technology and its effectiveness in solving a crisis plays a major role. The "not in my backyard" perception of landfills applies to the case of decentralized treatment facilities as well. It is believed that a property located near a treatment facility may have odour issues and hence may lower the value of the property with time. In order to address this issue, the authorities should draft laws and frameworks in such a way that in case of any odour or any form of the nuisance that has occurred due to any unforeseen reason, the residents in the surrounding area shall be heavily compensated. This step can have a significant impact on an individual's perception.

The institutional and legal frameworks that support or encourage the use of decentralized systems are lacking. The adoption of the systems in developing nations can be hampered by several variables, some of which include public engagement and awareness; programme assessment, inspection, and monitoring; system selection and design; institutional support; and financing. Adherence to the regulatory requirements for treated effluent is connected to the proof of an appropriate treatment system's efficacy in order for decentralized wastewater technologies to be widely used (Muzioreva et al., 2022).

Because it is challenging to properly maintain many treatment units, there is an increased danger of public health issues. Construction of the treatment, storage, or disposal units, along with an extra network for the reuse of grey water, raises the cost of the structure and is frequently not compensated by the cost of water saved.

7.10 CONCLUSIONS AND PERSPECTIVES

Decentralization has long been considered a desirable treatment strategy, and advancements in engineering and technology have promoted it to being widely used. The circular economy is boosted by decentralized wastewater treatment facilities that allow for the recovery and reuse of nutrients and water. Regardless of the benefits and drawbacks of decentralized systems, it is imperative for public acceptance that people feel that the current water management system must be altered in some way. If the general public does not prioritize environmental issues, strategies to solve the problem of water stress are likely to receive little or no support. Decentralized treatment and the need for its application must be conveyed to people in order to gain public acceptance. The centralized treatment is not economical due to the transportation of waste from source of generation to treatment facility, leakages, and regular maintenance. Instead, a decentralized facility provides many advantages, for example, on-site treatment; appropriate technology can be adopted for different types of waste and reuse of treated water. The decentralized treatment technique is suitable for every application, including commercial and domestic applications such as hospitals, railway stations, airports, bus stations, etc., and at every individual household. This reduces water pollution and waterborne diseases. Decentralization allows maximum reuse of treated water. Various case studies suggest that people are not effectively using the decentralized treatment facilities. In many cases, people are not using regularly due to cost of operation and in such cases state or national bodies should provide effective policies in order to address the challenges faced by people. Awareness and education regarding the benefits of decentralization treatment must be conveyed to the people through community engagement. Developing trust and confidence among the people regarding the systems' reliability and safety is vital. Community involvement in decision making and developing an affordable treatment unit could lead to public acceptance. Sharing successful stories and demonstration of potential benefits could develop confidence among the people. Effective application of decentralized wastewater treatment with appropriate guidelines will gain public acceptance and reduce water pollution and environmental pollution, and improve public health. Creating systems that are more environmentally friendly, socially responsible, and financially stable than traditional, centrally controlled systems is the aim of the decentralization method. But, developing nations need to have appropriate rules and institutional frameworks that incorporate sanitation policies that are consistent with decentralized technologies to a larger extent for the effective implementation or acceptance of decentralized wastewater systems.

REFERENCES

Asian Development Bank Institute (ADBI). Public–Private Partnerships for Wastewater Treatment in Rural Areas: Case Study of Changshu, People's Republic of China. (2022). https://www.adb.org/sites/default/files/publication/798661/adbi-case-study-2022-1.pdf.

Bernal, D., I. Restrepo, and S Grueso-Casquete. "Key criteria for considering decentralization in municipal wastewater management." *Heliyon* 7, no. 3 (2021): e06375.

Brunner, Norbert, Markus Starkl, Absar A Kazmi, Alvaro Real, Nitin Jain, and Vijay Mishra. "Affordability of decentralized wastewater systems: A case study in integrated planning from INDIA." *Water* 10, no. 11 (2018): 1644.

Chirisa, Innocent, Elmond Bandauko, Abraham Matamanda, and Gladys Mandisvika. "Decentralized domestic wastewater systems in developing countries: The case study of Harare (Zimbabwe)." *Applied Water Science* 7, no. 3 (2017): 1069–1078.

Ferreira, Marcella Moretti, Fabiana Alves Fiore, Alexandre Saron, and Gustavo Henrique Ribeiro da Silva. "Systematic review of the last 20 years of research on decentralized domestic wastewater treatment in Brazil: State of the art and potentials." *Water Science and Technology* 84, no. 12 (2021): 3469–3488.

Gómez-Román, C., L. Lima, S. Vila-Tojo, A. Correa-Chica, J. Lema, and J. M Sabucedo. "Who cares?": The acceptance of decentralized wastewater systems in regions without water problems. *International Journal of Environmental Research and Public Health* 17, no. 23 (2020): 9060.

Gómez-Román, C., J. M. Sabucedo, M. Alzate, and B Medina. "Environmental concern priming and social acceptance of sustainable technologies: The case of decentralized wastewater treatment systems." *Frontiers in Psychology* 12 (2021): 647406.

Jung, Youngmee Tiffany, N. C. Narayanan, and Yu-Ling Cheng. "Cost comparison of centralized and decentralized wastewater management systems using optimization model." *Journal of Environmental Management* 213 (2018): 90–97.

Kuttuva, Pradeep, Sharachchandra Lele and Gara Villalba. "Decentralized wastewater systems in Bengaluru, India: Success or failure?." *Water Economics and Policy* 04 (2018): 1650043. doi: 10.1142/S2382624X16500430.

Muzioreva, Happison, Trynos Gumbo, Neema Kavishe, Thembani Moyo, and Innocent Musonda. "Decentralized wastewater system practices in developing countries: A systematic review." *Utilities Policy* 79 (2022): 101442.

Nansubuga, Irene, Noble Banadda, Willy Verstraete, and Korneel Rabaey. "A review of sustainable sanitation systems in Africa." *Reviews in Environmental Science and Bio/Technology* 15, no. 3 (2016): 465–478.

Prateep Na Talang, Rutjaya, Sanya Sirivithayapakorn, and Sucheela Polruang. "Life cycle impact assessment and life cycle cost assessment for centralized and decentralized wastewater treatment plants in Thailand." *Scientific Reports* 12, no. 1 (2022): 1–13.

Roefs, I., B. Meulman, J. Vreeburg, and M Spiller. Centralised, decentralised or hybrid sanitation systems? Economic evaluation under urban development uncertainty and phased expansion. *Water Res*earch 109 (2017): 274–286. doi: 10.1016/j.watres.2016.11.051.

United Nations Economic and Social Commission for Asia and the Pacific (UNESCAP),United Nations Human Settlements Programme (UN-Habitat) and Asian Institute of Technology (AIT). (2015). Policy Guidance Manual on Wastewater Management with a Special Emphasis on Decentralized Wastewater Treatment Systems. Bangkok, Thailand.

Valentin, V. and S. M Bogus. "Assessing the link between public opinion and social sustainability in building and infrastructure projects." *Journal of Green Build*ing 10 (2015): 177–190. DOI: 10.3992/jgb.10.3.177

Wilson, T. D., D. S. Dunn, D. Kraft, and D. J Lisle. (1989). "Introspection, evaluation change, and evaluation-behavior consistency: The disruptive effects of explaining why we feel the way we do," in *Advances in Experimental Social Psychology*. Vol. 22. ed. L. Berkowitz (New York: Academic Press), 287–343.

8 Life Cycle Environmental Assessment of Decentralized Sanitation Systems

Pablo Gortáres-Moroyoqui[a], Luis Humberto Álvarez-Valencia[b], Gabriela Ulloa Mercado[a], Pasiano Rivas-García[c], Luis Alonso Leyva-Soto[a,d], Denisse Serrano-Palacios[e], Ana María Rentería-Mexía[a], Edna Rosalba Meza-Escalante[e], and Lourdes Mariana Díaz-Tenorio[a]

[a] Departamento de Biotecnología y Ciencias Alimentarias, Instituto Tecnológico de Sonora, Sonora, México

[b] Departamento de Ciencias del Agronómicas y Veterinarias, Instituto Tecnológico de Sonora, Sonora, México

[c] Facultad de Ciencias Químicas, Universidad Autónoma de Nuevo León, México

[d] Consejo Nacional de Ciencia y Tecnología (CONACYT), México

[e] Departamento de Ciencias del Agua y Medio Ambiente, Instituto Tecnológico de Sonora, Sonora, México

8.1 INTRODUCTION

Nowadays, everyone is at least partially aware of the situation around freshwater, concerning its scarcity, pollution, and high demand due to the fast-growing population. According to the United Nations World Water Development Report (2017), water is mainly used for agriculture. The freshwater withdrawals in the world are approximately 3,928 km^3 per year, out of which 44% is used by agriculture. However, 56% is released as industrial and municipal wastewater and agricultural drainage water (Aquastat, 2016). Figure 8.1 schematizes the uses of freshwater in the world.

The UN WWAP report declares that the practice of throwing waste (including wastewater) in water reservoirs like lakes, lagoons, rivers, and the sea has been decreasing in the last years, though it is still essential. Worldwide, the total amount of accessible and reliable water is about 4,200 billion m^3; however, the water demand in the year 2030 will be 6,900 billion m^3, representing an increment of 64% in Jabri et al., (2020). In this sense, the efforts to recycle wastewater face water scarcity, using decentralized systems as a promising alternative (Jabri et al., 2020). Singh and Colosi (2020) propose using brackish water as a drinking water supply or to support natural ecosystems. They assert that an appropriate desalination technology can use treated water as drinking water. Many examples of water scarcity and the need for better operational systems for reuse exist in the scientific literature as the importance of wastewater treatments, including the decentralized systems. However, in reality, that effort is still insufficient. For example, according to Bajpai et al. (2019), 2.3 billion people require proper sanitation systems. Then, the situation for water supply is critical, and taking care of, using correctly, treating, and reusing the water and, consequently, the wastewater is vital.

DOI: 10.1201/9781003376415-10

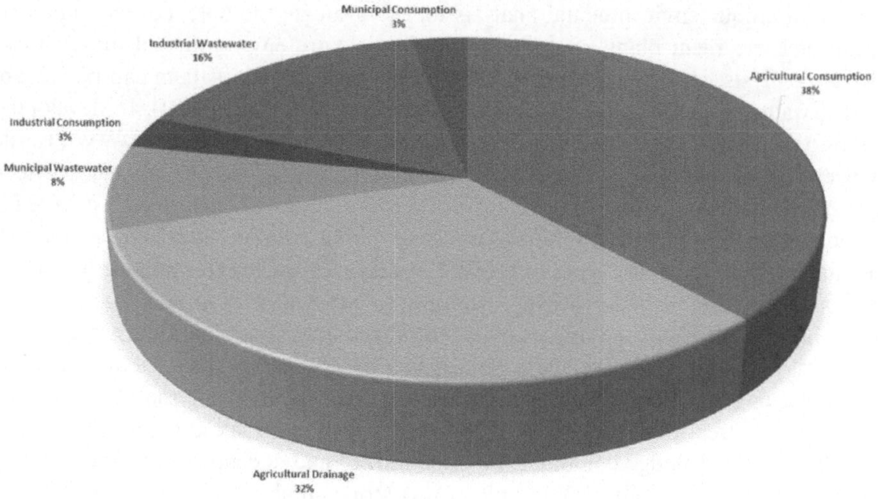

FIGURE 8.1 Withdrawals of freshwater in the world (data from WWAP, 2017).

8.2 LCA STUDIES ON DECENTRALIZED TREATMENT IN DEVELOPING COUNTRIES

The increased awareness of the importance of environmental protection and the possible impacts associated with products, both manufactured and consumed, has led to more interest in developing methods to better understand and address these impacts. Life cycle assessment (LCA) is a technique for this purpose. LCA can assist in (1) identifying opportunities to improve the environmental performance of products at various points in their life cycle; (2) informing decision-makers in industry, government, or non-government organizations; (3) the selection of relevant indicators of environmental performance, including measurement techniques; and (4) marketing (ISO, 2022). LCA is a tool that helps evaluate the environmental load (and, as a consequence, its impact) throughout its life cycle. However, this tool has been used for decades with different names (Parra-Saldivar et al., 2020).

In environmental topics, LCA means that the study includes all possible impacts, from the production to the final disposal; vulgarly, the tool is named from the cradle to the grave (Parra-Saldivar et al., 2020). The method can be used for products, substances, systems, and equipment. Its systematic evaluation approach consists of four phases: goal and scope definition, inventory analysis, impact assessment, and interpretation (Romeiko, 2020); however, it is necessary to conduct more studies on decentralized systems. Besides, it is not easy to define a centralized and a decentralized system; in fact, most papers need to define the concepts clearly, and others state a straightforward definition. For example, the definition in Ntostoglou et al. (2021) is that a decentralized system consists of relatively small-scale facilities usually located a short distance from waste sources and end users.

8.2.1 Decentralized Wastewater Treatment Plants

In developing countries, the causes and consequences for the necessity of good wastewater treatment are various; they are focused mainly on the fast growth of the population. This results in generation of different approaches or frameworks to solve the problem. These frameworks can be chosen using several criteria – from cost to impact, including the environment, economy, society, and aggregation – as pillars of sustainability (Starkl et al., 2022).

LCA is not a unique environmental analysis tool to evaluate the potential environmental risks from wastewater treatment plants. Some of analysis tools are environmental impact assessment, material flow analysis, and strategic environmental assessment; all of them can be valuable tools when a wastewater treatment plan is designed (Dewalkar & Shastri, 2021). In developing countries, studies on the impact assessment of decentralized wastewater treatment (DWWT) systems are scarce; however, in the last year, these have been increasing, mainly because of the cost-effectiveness they represent (Ntostoglou et al., 2021). Kasora and Mourad (2018) performed a detailed study to evaluate the sustainability of the operating collective, public, semi-centralized sewage systems in Kigali, Rwanda. There are many types of DWWT systems for wastewater treatment; one is anaerobic digestion, a very established process. According to Ntostoglou et al. (2021), it is necessary to enhance scientific evidence of the impact of decentralized anaerobic digestion systems in different urban contexts. However, there are some occasions when decentralized systems are not the best option. Khattiyavong and Lee (2019) conducted some experiments using the LCA tool to identify a suitable system to treat the wastewater of Vientiane. They used six alternatives based on trickling filters and activated sludge combined with centralized and decentralized systems. They found that the centralized system with trickling filters was more suitable than the decentralized system, according to environmental impacts such as land requirements, CO_2 emissions, power demand, and sludge production.

On the other hand, Arias et al. (2019) evaluated a decentralized membrane bioreactor installed in Turkey, using climate change and eutrophication potential as environmental indicators. They experimented with different retention times in the reactor and the influence of impacts associated with the construction phase. They found that the most significant impacts are the electricity consumption in the operational phase and the infrastructure in the construction phase. In addition, Singh et al. (2019) evaluated six DWWT systems installed in India, concluding that these technologies are good alternatives for treating wastewater, with negligible energy and chemical consumption. Estévez et al. (2022) recently published an interesting paper comparing the environmental impacts of centralized and decentralized systems in remote and small cities. In that study, the decentralized system recovered energy and nutrients from black water, improving the environment. DWWT systems are good emerging alternatives to more ancient centralized systems, and LCA is a suitable tool to evaluate all environmental impacts of these plants "from cradle to grave".

8.3 LCA STUDIES ON DECENTRALIZED TREATMENT IN DEVELOPED COUNTRIES

Large human settlements are localized sites where a vast amount of consumable goods is concentrated for the benefit of society. One means of transporting the losses and waste of these goods is through water sanitization systems, which have historically aimed to reduce the environmental burdens of these effluents, through water treatment systems, before being deposited in water bodies. In recent decades, the trend has been to design waste treatment systems that, in addition to being efficient in reducing environmental loads, allow the return of nutrients to agricultural fields of intensive food production, generate fuels without the need for fossil resources, and consider the generation of other valuable products. This archetype of sanitation systems contributes to a circular economy scheme. It is in line with the 10th principle of green engineering: The design of products, processes, and systems must include integration and interconnectivity with available energy and material flows (Jiménez-González & Constable, 2011). The decentralized treatment approach is in line with the philosophy and guidelines of the circular economy, as it allows the use of reclaimed water, bioenergy, and biofertilizers under safe and environmentally friendly conditions (Estévez et al., 2022).

A wastewater treatment system with co-product recovery sub-processes may have various design options depending on many factors. Tchobanoglous et al. (2003) identified 23 essential factors to consider during the design/selection of wastewater treatment processes. Compared to centralized systems, DWWT systems allow for more specific and pertinent inclusion of design factors (Gallagher & Gill, 2021). These factors must take into account critical criteria such as the characteristics of the influents and their variability; the consumption of energy and materials in operation; the emission of environmental loads during the treatment process; generation of waste and co-products with added value, investment, maintenance, and operation costs; possible income from the sale of co-products and those related to the mitigation of greenhouse gases such as carbon credits, optimal geographic location strategies, and land use. Figure 8.2 presents an evaluation approach that should be considered for the design of sustainable DWWT systems. In these schemes, applying the LCA is fundamental, not only by implementing it in a conventional way (with an environmental approach) but also by considering monetary and social aspects through the methodologies of life cycle costing and social LCA, respectively.

Large sanitation systems have come under increasing scrutiny regarding their social, economic, and environmental sustainability, with studies comparing centralized and decentralized water infrastructure (Arora et al., 2015; Coutard & Rutherford, 2015; Eggimann et al., 2016). In addition, the design and evaluation of water infrastructure have evolved from a technical approach to a more socio-technical one, considering changes in social dynamics, technology, how to be more friendly with environment or mother nature (Obertreis et al., 2016). These points have led to increased interest in the sustainability of DWWT systems that incorporate advances in treatment technology to protect against environmental contamination (Capodaglio et al., 2017; Zang et al., 2021; Hacker & Binz, 2021).

In developed countries, DWWT systems have been more successful in rural areas (Chirisa et al., 2017; Poustie et al., 2020; Zang et al., 2021). Reusing treated water can represent an attractive resource recovery option for small- and medium-sized agricultural areas. It not only meets irrigation needs but also provides a potential source of nutrients for crops (Poustie et al., 2020). The use of water treated by DWWT systems has become a reality. In the Mediterranean regions such as Murcia (Spain), more than 100 mm^3/year are reused for agriculture. In rural Milan (Italy), high-quality treated wastewater is used for agricultural irrigation (SUWANU Europe, 2019). California and Florida have been irrigating crops with treated water in the United States for decades (Parsons,

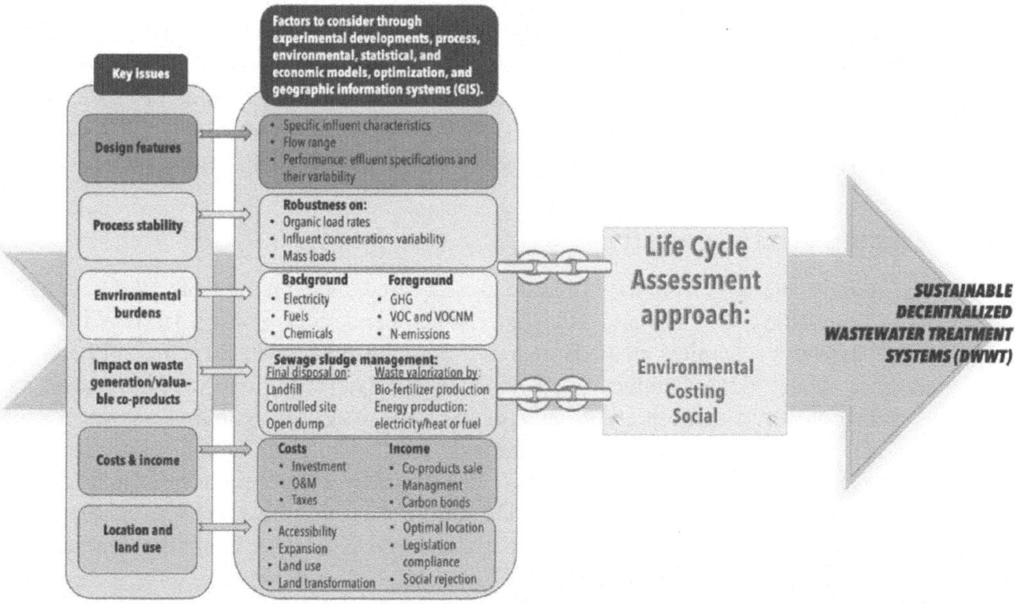

FIGURE 8.2 Holistic evaluation approach for sustainable decentralized wastewater treatment (DWWT) systems.

2018; Sheikh et al., 2018). DWWT systems present a viable solution from the economic and environmental point of view; they have low construction and operation costs and can offer partial solutions to reduce water stress in agricultural areas. However, they often face legal obstacles to operating effectively (Dubber & Gill, 2014).

LCA and life cycle costing can provide information on the environmental and economic performance of DWWT (Corominas et al., 2020). It can subsequently inform decision-making in designing and selecting specific treatment processes. On one hand, centralized WWT (CWWT) systems have high greenhouse gas emissions due to the high energy and material consumption during the management process (Gallagher & Gill, 2021). On the other hand, CWWT can incorporate sewage sludge management processes through anaerobic digestion, which mitigates greenhouse gas emissions in two ways: (1) avoiding emissions from sludge management in landfills and (2) replacing conventional energies (commonly based on fossil resources) with bioenergy and chemical fertilizers with biofertilizers. For their part, DWWT systems are commonly based on the segregation of three types of effluents: black water generated in toilets; gray water from laundry, showers, or dishwashers; and kitchen waste (Wielemaker et al., 2018). Black water can be treated through different types of anaerobic digesters, which, if properly designed, considering variable co-digestion schemes of optimal operation and design, help improve the systems' environmental and economic profiles (Miramontes-Martínez et al., 2022). At the same time, the management of gray water allows its reuse to recharge toilets and agricultural irrigation water (Skambraks et al., 2017).

The search for solutions in DWWT systems is a complex and multidimensional task. It requires studies that help decision-makers choose the best alternative, usually under a limited budget. In this framework, the LCA and life cycle costing methodologies play a fundamental role, allowing the identification of environmental hotspots, proposing strategies with a smaller environmental footprint, knowing which activities consume the most significant part of the budget, and how different alternatives for co-products valorization (such as biogas, biofertilizers, or simply water for agricultural irrigation) provide DWWT schemes with greater sustainability.

8.4 TECHNICAL PARAMETERS

Many studies on DWWT systems were considered to know the typical technical parameters for their control and operation. As a result, 236 DWWT systems were identified, and for the different parameters analyzed, such as biochemical oxygen demand (BOD), chemical oxygen demand (COD), total suspended solids (TSS), total nitrogen (TN), and total phosphorus (TP), the sample size achieved up to 111 values, which were taken from the different literature used in this section. In addition, the type of wastewater treated and energy consumption were also considered.

8.4.1 Treatment Systems and Size

Most of the systems analyzed from the literature consist of biological treatments (88%), with 57% operating under aerobic conditions and 31% under anaerobic systems. Only 24 DWWT systems were operated by physicochemical methods such as evapotranspiration (Paulo et al., 2019), adsorption (Rodríguez et al., 2021), disinfection (Slompo & da Silva, 2019), and oxidation-adsorption processes (Zaharia, 2017); that represents 12% of the total treatment systems analyzed. Among the 236 DWWT systems analyzed, the most commonly used technologies are the constructed wetlands (CW, 24%), anaerobic baffled reactor (ABR, 12%), up-flow anaerobic sludge blanket reactors (UASB, 9.7%), and anaerobic filters (AF, 6.8%). The fluidized aerobic reactor (FAR, 0.5%), rotating biological contactor (RBC, 0.9%), and oxidation-adsorption (OX-AD, 0.5%) are the treatment systems with the lowest percentage of use (Figure 8.3).

The CWWT are aerobic systems widely used as an alternative treatment to conventional systems such as activated sludge, which exhibit high treatment efficiency for organics and suspended solids. Depending on the configuration, CW can remove nutrients (Vymazal, 2019).

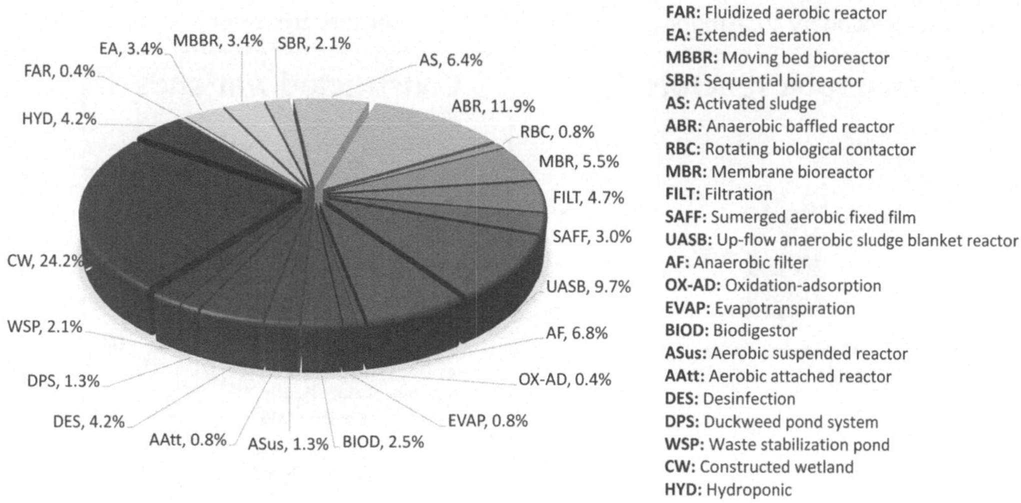

FIGURE 8.3 Different decentralized sanitation systems analyzed from the literature (Singh et al., 2015; Alvarado et al., 2017; Singh & Kazmi, 2018; Bajpai et al., 2019; Magwaza et al., 2020; Ferreira et al., 2021; Fernández del Castillo et al., 2022; Zaharia 2017; Singh et al., 2019).

These systems are soil or gravel-based systems planted with *T. latifolia* and/or *P. australis*, focused on treating urban runoff, domestic, mine drainage, and industrial wastewater (Cooper et al., 1996; Mays & Edwards, 2001; Scholz et al., 2007; Kadlec & Wallace, 2008). On the other hand, the ABR, UASB, and AF are anaerobic technologies with different reactor configurations designed to treat wastewater with high organic matter content. These systems offer advantages over aerobic systems, such as lower nutrient requirement, less surplus sludge production, small reactor volume, and energy recovery via methane production (Ghangrekar & Behera, 2014). Despite the high efficiency of anaerobic systems, their effluents often have a remnant of organic matter and nutrients that should be treated in a tertiary unit such as CW. The combination of anaerobic systems and CW have been used as an alternative in decentralized systems to overcome their disadvantages and to improve wastewater treatment (Liu et al., 2015). Fernández del Castillo et al. (2022) summarize different combination strategies of anaerobic systems with CW (Figure 8.4). They indicated that combined systems represent a suitable alternative for DWWT systems due to the efficient removal of organic matter, solids, and nutrients, to meet water quality standards.

The size of DWWT systems was analyzed according to the installed capacity in terms of flow rate (m^3/day) and population served, resulting in 67 and 40 plants, respectively (Figure 8.5). The minor treatment systems identified were (1) evapotranspiration plants to treat domestic wastewater for two people in a rural area (Paulo et al., 2019); (2) UASB + AF with a capacity of 0.3 m^3/day to treat domestic wastewater for three people in a rural area (Tavarez et al., 2020); (3) combined systems with anaerobic reactors and CWWT for domestic wastewater in a rural area for four to six people (Lutterbeck et al., 2018). The highest treatment system is an oxidation ditch operated with activated sludge with a capacity of 25,740 m^3/day (USEPA, 2000). Whereas considering the population served, the highest treatment systems analyzed were to treat effluents in communities of 10,000 and 5,000 inhabitants, operating by the duckweed pond system (Nhapi et al., 2003) and anaerobic baffled reactor coupled to CW (A. Singh et al., 2019). According to the flow rate, 34 DWWT systems have an installed capacity from 0.3 to <100 m^3/day, 20 systems from 100 to <1,000 m^3/day, nine systems from 1,000 to <10,000 m^3/day, and four systems from 10,000 to 26,000 m^3/day.

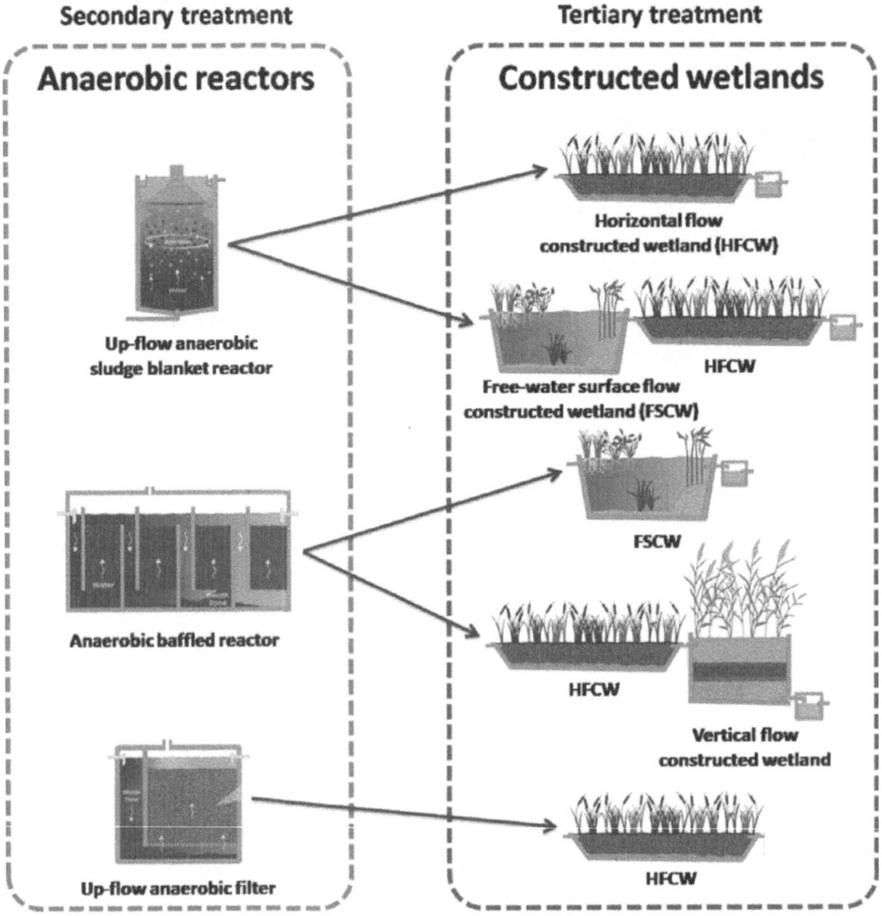

FIGURE 8.4 Combined decentralized systems applying anaerobic reactor coupled to constructed wetlands (modified from Fernández del Castillo et al., 2022).

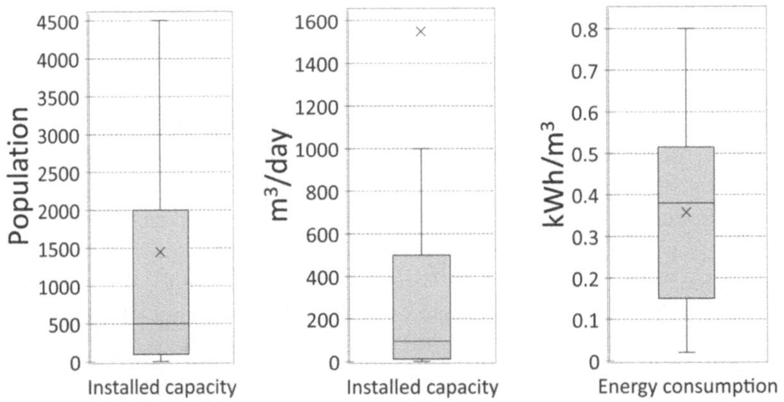

FIGURE 8.5 Installed capacity determined by the population served and flow rate (m^3/day) and energy consumption of decentralized sanitation systems.

8.4.2 Type of Wastewaters

DWWT systems analyzed focus on treating industrial, domestic, rural, hospital, and residential wastewaters. The most common type of wastewater is from rural areas (55%), followed by domestic effluents (22%), in which different treatment technologies are used. The industrial sector contributes 13% of the total type of effluents in the systems analyzed. Examples of small industry systems are the brewery industry with the hydroponic system (Worku et al., 2018), the wine industry with an anaerobic reactor coupled to CW (Serrano et al., 2011), and the tannery industry with different configurations of CW (Saeed et al., 2012). Effluents from hospitals represent 7% of the treatment systems analyzed; between them, the ABR (Simon Ndibnuh et al., 2018), different configurations of CW (Auvinen et al., 2017; Vo et al., 2019), moving bed bioreactor (Ooi et al., 2018), and membrane bioreactor (Beier et al., 2011) are some examples of DWWT systems used.

8.4.3 Treatment Efficiency

BOD, COD, TSS, ammonium, phosphate, TN, and TP are the most common parameters evaluated in the analyzed DWWT systems, but the main parameters are TSS, BOD, and COD. The removal efficiencies vary considerably for the parameters analyzed between the different systems, especially for ammonium, TN, and TP, representing removal efficiencies from 2.8% to 99% (Figure 8.6). The highest BOD removal efficiencies (≥90%) were observed in 43 decentralized systems; some examples are the ABR (Singh et al., 2019), CW (Tsihrintzis et al., 2010), and MBR (Singh & Kazmi, 2018) systems to treat domestic wastewater. Meanwhile, the lowest values (≤50%) were observed by six treatment plants, one of them being the ABR system to treat domestic wastewater (Singh et al., 2019). Removal efficiencies ≥90% for COD and TSS were achieved by 39 and 43 treatment systems, respectively. Considering all data collected, the removal efficiencies (%) were BOD, 80 ± 18 (n = 99); COD, 74 ± 23 (n = 111); TSS, 78 ± 21 (n = 100); ammonium, 66 ± 32 (n = 56); phosphate, 75 ± 20 (n = 7); TN, 57 ± 25 (n = 73); and TP, 51 ± 27 (n = 76).

8.4.4 Energy Consumption

Few studies have documented energy consumption. Among 236 DWWT systems analyzed, only 17 studies include energy consumption ranging from 0.02 to 0.8 kWh/m^3, with an average of 0.35 kWh/m^3 (Figure 8.5). The lowest energy consumption occurred in anaerobic reactors; meanwhile, the highest consumption was observed in MBBR systems (Singh & Kazmi, 2018). The typical range of energy consumption in DWWT systems is from 0.1 to 1.5 kWh/m^3 (Singh et al., 2019).

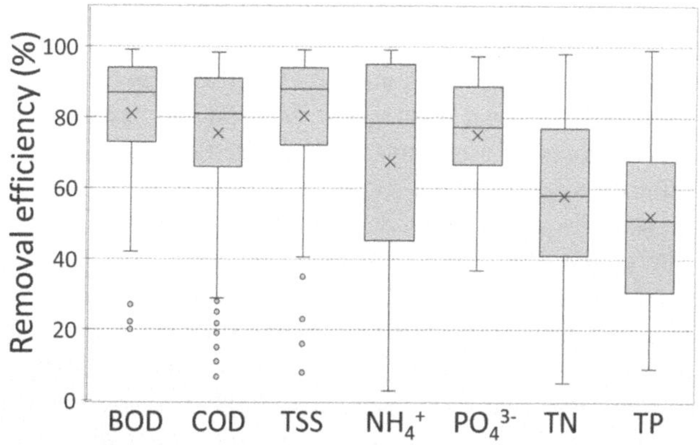

FIGURE 8.6 Removal efficiencies of the main parameters evaluated in different decentralized sanitation systems as those considered in Figure 8.3.

8.5 ENVIRONMENTAL BENEFITS

Up to this point, LCA studies of decentralized treatment systems in developing and developed countries have been analyzed; likewise, the technical parameters that must be considered to carry out an LCA of decentralized sanitation systems have been studied. As a result, LCA studies of decentralized sanitation systems can achieve to define the effects in some categories established in ISO 19115-Topic Category: farming, biota, boundaries, climatology-meteorology-atmosphere, economy, elevation, environment (including environmental resources, protection, and conservation; examples: environmental pollution, waste storage and treatment, environmental impact assessment, monitoring environmental risk, nature reserves, landscape), geoscientific-information, health, imagery-base-maps-earth-cover, intelligence military, inland waters, oceans, planning cadastre, society, structure, transportation, utilities of communication (ISO 19115).

In this sense, the environmental benefits detected through LCA studies of decentralized sanitation systems can be reflected in pollution reduction, ozone layer protection, and global warming prevention; through less emission of greenhouse gases, mainly CO_2, when renewable or reusable materials are used for the construction of sanitation systems instead of nonrenewable materials, this leads to a decrease in the use of raw materials. It assures us of sustainability and the protection of natural resources. To exemplify this situation, Pryce et al. (2021) tested five primary materials (stainless steel, mild steel, glass fiber reinforced polymer, high-density polyethylene, and reinforced concrete cement) in the construction of small wastewater treatment plants. When the LCA was applied to the construction process of the plant, they found that stainless steel was the material with the higher impact, mainly in the health category, and reinforced concrete cement incurred the least environmental cost. The main environmental impacts were global warming, nonrenewable energy, and terrestrial ecotoxicity. Also, Kobayashi et al. (2020) performed the LCA on gray water treatment systems with reuse at different scales in cold regions. They used constructed wetland and membrane bioreactors (MBR) at three different scales. The main environmental benefit was that large scales reduced global warming, eutrophication potential, and human health-carcinogenic potential. The wetland was the scenario with the higher environmental benefits when used in small communities or neighborhoods. MBR is best used in communities where a large amount of gray water is reused.

On the other hand, the inventory level is the key to improving the quality of the LCA with the consequent more accurate calculation of direct greenhouse gas emissions. Suppose the description of supply inputs, their origins, and technical characteristics are more precise. In that case, the inventory includes all materials and resources to approach through LCA of the treatment system to actual emissions and environmental effects. It is necessary to define the environmental benefits or adverse effects. Eutrophication and global warming are the two most commonly assessed impacts (Gallego-Schmid & Tarpani, 2019).

Table 8.1 shows some LCA studies in treatment systems where different factors were evaluated, such as scales, different construction materials of the systems, processes with and without reuse of treated water, various types of treated waste, and different treatment configurations. As a result of the LCA studies, the factors that cause a less negative impact on the environment or more significant environmental benefits were found. For example, decentralized systems have 20% less impact on the environment than centralized systems. In other work, by changing the material of construction of the treatment system from stainless steel to high-density polyethylene, the emission of CO_2 is reduced from 8.7 tons of CO_2 through fossil fuel use to less than 0.73 tons CO_2 in total, respectively, contributing less to climate change (Pryce et al., 2021). Also, CO_2 emissions can be reduced by up to 70% when recycled material is used instead of virgin material. In this case, this could be achieved by having a good metal collection and recycling system so that the recycled material necessary for the construction of the treatment systems is available (Johnson et al., 2008). On the other hand, Tian et al. (2021) compared the sustainability of food waste anaerobic digestion (with different biogas applications) versus incineration. They found that concerning the global warming potential, all scenarios achieved negative results, which means the analyzed technologies contributed to slowing down climate change.

TABLE 8.1
Diversity of Sanitation System, Main Studied Scenarios, and the Most Important Environmental Benefits Obtained in the LCA Studies

Decentralized Treatment System	Studied Scenarios	Environmental Benefits Findings	Reference
Non-potable reuse application	Recycling treated gray water	Reduction of between 3.9% and 17% in five impact categories	Masmoudi Jabri et al., 2020
Seawater desalination options	Centralized vs. decentralized scenarios	Environmental impacts 20% lower for decentralized scenarios than for centralized	Shahabi et al., 2015
Gray water treatment systems with reuse at different scales in cold regions	Constructed wetlands (CW) vs. membrane bioreactors (MBR) in three different scales: (1) 3,500 persons equivalent [PE]), (2) 350 PE, and (3) up to 5 PE	Larger scales reduced global warming potential (GWP), eutrophication potential (EUP), and human health–carcinogenic potential (HHCP).	Kobayashi et al., 2020
Food waste anaerobic digestion with different downstream biogas utilization	Anaerobic digestion (AD) vs. incineration	AD scenarios were generally preferable compared to incineration. Maximizing the methane yield and minimizing the food waste collection radius were also identified as essential measures to improve the overall environmental performance.	Tian et al., 2021
Construction of small wastewater treatment plants	Investigated five primary materials (stainless steel, mild steel, glass fiber reinforced polymer (GFRP), high-density polyethylene, and reinforced concrete cement)	Stainless steel demonstrated a substantially higher impact. The main environmental impacts were global warming, nonrenewable energy, and terrestrial ecotoxicity.	Pryce et al., 2021
Alternatives for improvement of wastewater treatment plants in developing countries	Current conventional WWTPs in developing countries by adding tertiary treatment and/or anaerobic digestion of sludge	The scenario of WWTPs + the addition of anaerobic digestion of sludge and tertiary treatment stage attained maximum environmental benefits for all categories due to the energy saving and the prospect of water reuse.	Awad et al., 2019
Polish individual wastewater treatment	Stage of construction, use, and end-of-life	The primary GHG were CO_2, CH_4, and nitrous oxide, in the same abundance order.	Burchart-Korol & Zawartka, 2019
Wastewater treatment systems for domestic homes	Evaluated six unique life cycle environmental impacts for different DWTTS configurations of five conventional septic tank systems, four packaged treatment units, and a willow evapotranspiration system	Similar freshwater eutrophication (FE), dissipated water (DW), and mineral and metal (MM) burdens were noted between the packaged and conventional system configurations, with the packaged systems demonstrating significantly higher impacts of between 18% and 56% for climate change (CC), marine eutrophication (ME), and fossils (F).	Gallagher & Gill, 2021

Food waste anaerobic digestion followed by biogas used as cooking fuel showed the best environmental performance, with a net greenhouse gas emission reduction of up to 238 kg CO_2-eq/ton of disposed food waste. All anaerobic digestion analyzed showed a better environmental performance than incineration. In the same sense, Bertanza et al. (2017) compare conventional activated sludge and membrane bioreactor (MBR) processes for treating municipal wastewater. The results showed, in the category of environmental aspects, that both treatment systems had a similar effect on global warming potential, acidification potential, and eutrophication terrestrial. MBR harmed the eutrophication of freshwater, while conventional activated sludge had a positive effect; contrary, MBR has a positive environmental effect in the eutrophication of marine, photochemical oxidant, and formation potential. The general evaluation of the environmental footprint is neutral for the MBR system since it has a lower use of chemicals that compensates for the higher energy consumption.

Undoubtedly, LCA studies are beneficial to determine the environmental implications or effects of processes or products. It is a tool that helps us determine the possible strategies or scenarios to reduce the impacts or improve the environmental benefits of production or service processes. However, there are few or no studies of treatment plants that include the removal of micropollutants or emerging contaminants, including antibiotics; hormones; pharmaceutical products such as ibuprofen and naproxen; and personal care products such as fragrances, dyes, and synthetic antioxidants (Gallego-Schmid & Tarpani, 2019).

8.6 IMPROVING ENVIRONMENTAL IMPACTS

LCA studies must consider the potential environmental impacts of emissions and resource use. Global warming (climate change or carbon footprint), eutrophication, and acidification are common impact categories analyzed. In recent years, other categories, such as ecotoxicities, human toxicity, ozone depletion, water consumption, water depletion, fossil fuel or mineral depletion, energy consumption, energy demand, and land use have gained the interest of study (Gallego-Schmid & Tarpani, 2019). When the impact categories have been defined, the next step is classification, whereby system inputs and outputs that have been compiled are assigned qualitatively to one or more of the impact categories (McNamara et al., 2016).

LCA can reduce the environmental load of a process by opting for an alternative production, processing, and distribution pattern in a multifaceted manner (Parra-Saldivar et al., 2020). Therefore, applying LCA is a suitable decision-making approach before projecting the implementation stage to give valuable insights into possible environmental impact, identifying the most critical influencing parameters, or identifying the best configurations compared to other technologies (Kataki et al., 2021). Decentralized systems aim to be economically affordable, socially acceptable, and environmentally sustainable for excreta and wastewater management. These systems are typically used in sparse residencies and small communities of rural areas and are considered a more sustainable technology than centralized systems. Moreover, decentralized systems do not require energy, are expensive, or have complicated operations, contrary to centralized systems. (Lourenço & Nunes, 2020).

Decentralized systems can adapt quickly to local conditions in urban areas and expand their capacity following population growth. In addition, this approach facilitates reusing water and treatment byproducts, such as energy, sludge, and nutrients. At the same time, decentralized wastewater treatment systems can reduce the number of pollutants water bodies receive after exiting the plant, reducing its environmental impact on parameters such as diluted oxygen, chemical oxygen demand, and biochemical oxygen demand (Bernal et al., 2021).

Adopting decentralized systems for wastewater treatment has excellent potential for protecting hydrologic and soil system resources, mainly surface water and groundwater. The most useful decentralized systems are those of separated sources in which gray and black wastewater are separated for additional treatment (Shawkat Mizzouri et al., 2022). Mladenov et al. (2022) evaluated common persistent and removed compounds found in two centralized wastewater treatment plants in the United States and South Africa and one decentralized wastewater treatment system in South Africa. Overall, the results proved that removal efficiencies of chemicals were similar between the treatment plants when they were compared according to the number of chemical features detected in the influents and effluents of each treatment plant. Nevertheless, the treatment train for decentralized systems, which has more extended solids retention and hydraulic residence times than both of the centralized wastewater treatment plants and utilizes primarily anaerobic treatment processes, was able to remove 13 additional compounds and showed a more significant decrease in normalized peak areas compared to the centralized wastewater treatment plants. Benetto et al. (2009) stated that decentralized systems could reduce the contribution to ecosystem quality damage by more than 60% compared to conventional systems.

Gathering energy consumption data is crucial and needs to be accounted for with care in LCA studies (Corominas et al., 2020). Electrical consumption is another important example of an environmental impact that decentralized systems can reduce. Prateep Na Talang et al. (2022) argue that electrical consumption contributes to almost all environmental impacts in treatment scenarios evaluated in their research. Furthermore, most environmental impacts, including electrical consumption of decentralized treatment scenarios, are lower than those of centralized treatment scenarios. Due to water scarcity, technologies that can reduce water and energy demands will be considered an excellent value. Although direct non-potable reuse of gray water can be performed on-site, various states and municipalities may still restrict its use over human health concerns.

Nevertheless, some studies have shown that for gray water systems in general, MBRs are a viable cost-reducing option for cities looking to reduce water withdrawals and energy generation. Also, they can reduce domestic non-potable water demand and energy consumption from centralized systems without increasing the environmental impacts (Jeong et al., 2018). However, there is more knowledge to pursue regarding the environmental sustainability of decentralized wastewater treatment systems. Therefore, actions toward obtaining future knowledge must continue to minimize environmental impacts and achieve sustainable sanitation goals (Gallego-Schmid & Tarpani, 2019).

8.7 CONCLUSIONS AND PERSPECTIVES

Around 2.3 billion people require proper sanitation systems. Therefore, the current conditions for water supply are critical, and taking care of, using correctly, treating, and reusing the water and, consequently, the wastewater is vital. DWWT systems are an excellent emerging alternative to more ancient centralized systems. The LCA is a suitable tool to evaluate the environmental impacts of these systems from the cradle to the grave.

The most used parameters to evaluate the DWWT by the LCA are the type of treatment systems and size, installed capacity, characteristics of wastewater, treatment efficiency, and energy consumption. Regarding the treatment efficiency, in general, the performance of DWWT described in this chapter could be rated as acceptable considering the most used parameters (COD, BOD, TSS, NH_4^+, TN, and TP) to evaluate the efficiency of wastewater treatment plants. Between the DWWT analyzed, constructed wetlands, UASB reactor, anaerobic filters, and anaerobic baffled reactor are the most used systems. In addition, combined systems, such as anaerobic reactors coupled to constructed wetlands, represent a suitable alternative for DWWT systems due to the efficient removal of organic matter, solids, and nutrients, to meet water quality standards.

The LCA and life cycle costing methodologies play a fundamental role, allowing the identification of environmental hotspots, proposing strategies with a smaller environmental footprint, knowing which activities consume the most significant part of the budget, and which are different alternatives for co-product valorization. LCA studies are beneficial to determine the environmental implications or effects of processes or products. It is a tool that helps us determine the possible strategies or scenarios to reduce the impacts or improve the environmental benefits of production or service processes. However, there are few or no studies of treatment plants that include the removal of micropollutants or emerging contaminants. There is more knowledge to pursue regarding the environmental sustainability of DWWT systems. Therefore, actions toward obtaining future knowledge must continue to minimize environmental impacts and achieve sustainable sanitation goals.

REFERENCES

Alvarado, A., Larriva, J., Sánchez, E., Idrovo, D., & Cisneros, J. F (2017). Assessmenof DWWT systems in the rural area of Cuenca, Ecuador. *Water Practice and Technology*, *12*(1), 240–249. https://doi.org/10.2166/WPT.2017.027

Aquastat, F. A. O. (2016). Water withdrawal by sector, around 2006. Update November.

Arias, A., Vallina, I., Lorenzo, Y., Komesli, O. T., Katsou, E., Feijoo, G., & Moreira, M. T. (2019) Water Footprint of a Decentralized Wastewater Treatment Strategy Based on Membrane Technology. *Environmental water footprints, environmental footprints, and eco-design of products and processes*. S. S. Muthu editor. Book series, Springer (Publisher), Switzerland AG., 85–119. https://doi.org/10.1007/978-981-13-2508-3_4

Arora, M., Malano, H., Davidson, B., Nelson, R., & George, B. (2015). Interactions between centralized and decentralized water systems in urban context: A review. Wiley Interdisciplinary Reviews: *Water*, *2*(6), 623–634. https://doi.org/10.1002/wat2.1099

Auvinen, H., Havran, I., Hubau, L., Vanseveren, L., Gebhardt, W., Linnemann, V., Van Oirschot, D., Du Laing, G., & Rousseau, D. P. L. (2017). Removal of pharmaceuticals by a pilot aerated sub-surface flow constructed wetland treating municipal and hospital wastewater. *Ecological Engineering*, *100*, 157–164. https://doi.org/10.1016/J.ECOLENG.2016.12.031

Awad, H., Alalm, M. G., & El-Etriby, H. K. (2019). Environmental and cost life cycle assessment of different alternatives for improvement of wastewater treatment plants in developing countries. *Science of the Total Environment*, *660*, 57–68. https://doi.org/10.1016/j.scitotenv.2018.12.386

Bajpai, M., Katoch, S., Chaturvedi, N. K., Singh, S., Naveen, K., & Chaturvedi, K. (2019). Comparative study on decentralized treatment technologies for sewage and graywater reuse – A review. *Water Science and Technology*, *80*(11), 2091–2106. https://doi.org/10.2166/WST.2020.039

Beier, S., Cramer, C., Köster, S., Mauer, C., Palmowski, L., Schröder, H. F., & Pinnekamp, J. (2011). Full scale membrane bioreactor treatment of hospital wastewater as forerunner for hot-spot wastewater treatment solutions in high density urban areas. *Water Science and Technology*, *63*(1), 66–71. https://doi.org/10.2166/WST.2011.010

Benetto, E., Nguyen, D., Lohmann, T., Schmitt, B., & Schosseler, P. (2009). Life cycle assessment of ecological sanitation system for small-scale wastewater treatment. *Science of the Total Environment*, *407*(5), 1506–1516. https://doi.org/10.1016/j.scitotenv.2008.11.016

Bernal, D., Restrepo, I., & Grueso-Casquete, S. (2021). Key criteria for considering decentralization in municipal wastewater management. *Heliyon*, *7*(3). https://doi.org/10.1016/j.heliyon.2021.e06375

Bertanza, G., Canato, M., Laera, G., Vaccari, M., Svanström, M., & Heimersson, S. (2017). A comparison between two full-scale MBR and CAS municipal wastewater treatment plants: Techno-economic-environmental assessment. *Environmental Science and Pollution Research*, *24*(21), 17383–17393. https://doi.org/10.1007/s11356-017-9409-3

Burchart-Korol, D., & Zawartka, P. (2019). Determinants of environmental assessment of Polish individual wastewater treatment plants in a life cycle perspective. *Archives of Environmental Protection*, *45*(3). https://doi.org/10.24425/aep.2019.128640

Capodaglio, A. G., Callegari, A., Cecconet, D., & Molognoni, D. (2017). Sustainability of decentralized wastewater treatment technologies. *Water Practice and Technology*, *12*(2), 463–477. https://doi.org/10.2166/wpt.2017.055

Chirisa, I., Bandauko, E., Matamanda, A., & Mandisvika, G. (2017). Decentralized domestic wastewater systems in developing countries: The case study of Harare (Zimbabwe). *Applied Water Science*, *7*(3), 1069–1078. https://doi.org/10.1007/s13201-016-0377-4

Cooper, P. F., Job, G. D., Green, M. B., & Shutes, R. B. E. (1996). *Reed beds and constructed wetlands for wastewater treatment*. WRc Publications.

Corominas, L., Byrne, D. M., Guest, J. S., Hospido, A., Roux, P., Shaw, A., & Short, M. D. (2020). The application of life cycle assessment (LCA) to wastewater treatment: A best practice guide and critical review. *Water Research*, *184*, 116058. https://doi.org/10.1016/j.watres.2020.116058

Coutard, O., & Rutherford, J. (Eds.). (2015). *Beyond the networked city: Infrastructure reconfigurations and urban change in the North and South*. Routledge.

Dewalkar, S., & Shastri, S. (2021). Environmental impact assessment of wastewater treatment systems: State of the art. *International Journal of Advance and Innovative Research*, *8*, 307–214.

Dubber, D., & Gill, L (2014). Application of on-site wastewater treatment in Ireland and perspectives on its sustainability. *Sustainability*, *6*(3), 1623–1642. https://doi.org/10.3390/su6031623

Eggimann, S., Truffer, B., & Maurer, M. (2016). Economies of density for on-site waste water treatment. *Water Research*, *101*, 476–489. https://doi.org/10.1016/j.watres.2016.06.011

Estévez, S., González-García, S., Feijoo, G., & Moreira, M. T. (2022). How decentralized treatment can contribute to the symbiosis between environmental protection and resource recovery. *Science of the Total Environment*, *812*, 151485. https://doi.org/10.1016/j.scitotenv.2021.151485

Fernández del Castillo, A., Garibay, M. V., Senés-Guerrero, C., Orozco-Nunnelly, D. A., de Anda, J., & Gradilla-Hernández, M. S. (2022). A review of the sustainability of anaerobic reactors combined with constructed wetlands for decentralized wastewater treatment. *Journal of Cleaner Production*, *371*, 133428. https://doi.org/10.1016/J.JCLEPRO.2022.133428

Ferreira, M. M., Fiore, F. A., Saron, A., & da Silva, G. H. R. (2021). Systematic review of the last 20 years of research on decentralized domestic wastewater treatment in Brazil: State of the art and potentials. *Water Science and Technology*, *84*(12), 3469–3488. https://doi.org/10.2166/WST.2021.487

Gallagher, J., & Gill, L. W. (2021). The life cycle environmental performance of on-site or decentralised wastewater treatment systems for domestic homes. *Water*, *13*(18), 2542. https://doi.org/10.3390/w13182542

Gallego-Schmid, A., & Tarpani, R. R. Z. (2019). Life cycle assessment of wastewater treatment in developing countries: A review. *Water Research*, *153*, 63–79. https://doi.org/10.1016/j.watres.2019.01.010

Ghangrekar, M. M., & Behera, M. (2014). Suspended growth treatment processes. *Comprehensive Water Quality and Purification*, *3*, 74–89. https://doi.org/10.1016/B978-0-12-382182-9.00087-6

Hacker, M. E., & Binz, C. (2021). Institutional barriers to on-site alternative water systems: A conceptual framework and systematic analysis of the literature. *Environmental Science & Technology*, *55*(12), 8267–8277. https://doi.org/10.1021/acs.est.0c07947

Jabri, K., Nolde, E., Ciroth, A., & Bousselmi, L. (2020). Life cycle assessment of a decentralized greywater treatment alternative for non-potable reuse application. *International Journal of Environmental Science and Technology*, *17*, 433–444. https://doi.org/10.1007/s13762-019-02511-3

Jeong, H., Broesicke, O. A., Drew, B., & Crittenden, J. C. (2018). Life cycle assessment of small-scale greywater reclamation systems combined with conventional centralized water systems for the City of Atlanta, Georgia. *Journal of Cleaner Production*, *174*, 333–342. https://doi.org/10.1016/j.jclepro.2017.10.193

Jiménez-González, C., & Constable, D. J. (2011). *Green chemistry and engineering: A practical design approach*. John Wiley & Sons.

Johnson, J., Reck, B. K., Wang, T., & Graedel, T. E. (2008). The energy benefit of stainless steel recycling. *Energy Policy*, *36*(1), 181–192. https://doi.org/10.1016/j.enpol.2007.08.028

Kadlec, R. H., & Wallace, S. (2008). Treatment Wetlands. In *Treatment wetlands*. CRC Press. https://doi.org/10.1201/9781420012514

Kasora, A., & Mourad, K. (2018). Assessing the sustainability of decentralized wastewater treatment systems in Rwanda. *Sustainability*, *10*, 1–15. https://doi.org/10.3390/su10124617

Kataki, S., Chatterjee, S., Vairale, M. G., Sharma, S., Dwivedi, S. K., & Gupta, D. K. (2021). Constructed wetland, an eco-technology for wastewater treatment: A review on various aspects of microbial fuel cell integration, low-temperature strategies and life cycle impact of the technology. *Renewable and Sustainable Energy Reviews*, *148*(April), 111261. https://doi.org/10.1016/j.rser.2021.111261

Khattiyavong, C, & Lee, H (2019). Performance simulation and assessment of an appropriate wastewater treatment technology in A densely populated growing city in A developing country: A case study in Vientiane, Laos. *Water*, *11*, 1–15. https://doi.org/10.3390/w11051012

Kobayashi, Y., Ashbolt, N. J., Davies, E. G., & Liu, Y (2020). Life cycle assessment of decentralized greywater treatment systems with reuse at different scales in cold regions. *Environment International*, *134*, 105215. https://doi.org/10.1016/j.envint.2019.105215

Liu, R., Zhao, Y., Doherty, L., Hu, Y., & Hao, X (2015). A review of incorporation of constructed wetland with other treatment processes. *Chemical Engineering Journal*, *279*, 220–230. https://doi.org/10.1016/J.CEJ.2015.05.023

Lourenço, N., & Nunes, L. M. (2020). Review of dry and wet decentralized sanitation technologies for rural areas: Applicability, challenges, and opportunities. *Environmental Management*, *65*(5), 642–664. https://doi.org/10.1007/s00267-020-01268-7

Lutterbeck, C. A., Zerwes, F. V., Radtke, J. F., Köhler, A., Kist, L. T., & Machado, Ê. L (2018). Integrated system with constructed wetlands for the treatment of domestic wastewaters generated at a rural property – Evaluation of general parameters ecotoxicity and cytogenetics. *Ecological Engineering, 115*, 1–8. https://doi.org/10.1016/J.ECOLENG.2018.01.004

Magwaza, S. T., Magwaza, L. S., Odindo, A. O., & Mditshwa, A (2020). Hydroponic technology as decentralised system for domestic wastewater treatment and vegetable production in urban agriculture: A review. *Science of the Total Environment, 698*, 134154. https://doi.org/10.1016/J.SCITOTENV.2019.134154

Masmoudi Jabri, K., Nolde, E., Ciroth, A., & Bousselmi, L (2020). Life cycle assessment of a decentralized greywater treatment alternative for non-potable reuse application. *International Journal of Environmental Science and Technology, 17*(1), 433–444. https://doi.org/10.1007/s13762-019-02511-3

Mays, P. A., & Edwards, G. S (2001). Comparison of heavy metal accumulation in a natural wetland and constructed wetlands receiving acid mine drainage. *Ecological Engineering, 16*(4), 487–500. https://doi.org/10.1016/S0925-8574(00)00112-9

McNamara, G., Fitzsimons, L., Horrigan, M., Phelan, T., Delaure, Y., Corcoran, B., Doherty, E., & Clifford, E (2016). Life cycle assessment of wastewater treatment plants in Ireland. *Journal of Sustainable Development of Energy, Water and Environment Systems, 4*(3), 216–233. https://doi.org/10.13044/j.sdewes.2016.04.0018

Miramontes-Martínez, L. R., Rivas-García, P., Briones-Cristerna, R. A., Abel-Seabra, J. E., Padilla-Rivera, A., Botello-Álvarez, J. E., & Levasseur, A. (2022). Potential of electricity generation by organic wastes in Latin America: A techno-economic-environmental analysis. *Biomass Conversion and Biorefinery*, 1–12. https://doi.org/10.1007/s13399-022-03393-1

Mladenov, N., Dodder, N. G., Steinberg, L., Richardot, W., Johnson, J., Martincigh, B. S., Buckley, C., Lawrence, T., & Hoh, E (2022). Persistence and removal of trace organic compounds in centralized and decentralized wastewater treatment systems. *Chemosphere, 286*(P1), 131621. https://doi.org/10.1016/j.chemosphere.2021.131621

Nhapi, I., Dalu, J., Ndamba, J., Siebel, M., & Gijzen, H (2003). An evaluation of duckweed-based pond systems as an alternative option for decentralised treatment and reuse of wastewater in Zimbabwe. *Water Science & Technology, 48*(2), 323–330.

Ntostoglou, E., Khatiwara, D., & Martin, V (2021). The potential contribution of decentralized anaerobic digestion towards urban biowaste recovery systems: A scoping review. *Sustainability, 13*, 1–21. https://doi.org/10.3390/su132313435

Obertreis, J., Moss, T., Mollinga, P., & Bichsel, C. (2016). Water, infrastructure and political rule: Introduction to the special issue. *Water Alternatives, 9*(2), 168–181.

Ooi, G. T. H., Tang, K., Chhetri, R. K., Kaarsholm, K. M. S., Sundmark, K., Kragelund, C., Litty, K., Christensen, A., Lindholst, S., Sund, C., Christensson, M., Bester, K., & Andersen, H. R (2018). Biological removal of pharmaceuticals from hospital wastewater in a pilot-scale staged moving bed biofilm reactor (MBBR) utilising nitrifying and denitrifying processes. *Bioresource Technology, 267*, 677–687. https://doi.org/10.1016/J.BIORTECH.2018.07.077

Parra-Saldivar, R., Bilal, M., & Iqbal, H. M. N (2020). Life cycle assessment in wastewater treatment technology. *Current Opinion in Environmental Science and Health, 13*, 80–84. https://doi.org/10.1016/j.coesh.2019.12.003

Parsons, L. R (2018). Agricultural use of reclaimed water in Florida: Food for thought. *Journal of Contemporary Water Research & Education, 165*(1), 20–27. https://doi.org/10.1111/j.1936-704X.2018.03290.x

Paulo, P. L., Galbiati, A. F., Magalhães Filho, F. J. C., Bernardes, F. S., Carvalho, G. A., & Boncz, M. Á (2019). Evapotranspiration tank for the treatment, disposal and resource recovery of blackwater. *Resources, Conservation and Recycling, 147*, 61–66. https://doi.org/10.1016/J.RESCONREC.2019.04.025

Poustie, A., Yang, Y., Verburg, P., Pagilla, K., & Hanigan, D (2020). Reclaimed wastewater as a viable water source for agricultural irrigation: A review of food crop growth inhibition and promotion in the context of environmental change. *Science of the Total Environment, 739*, 139756. https://doi.org/10.1016/j.scitotenv.2020.139756

Prateep Na Talang, R., Sirivithayapakorn, S., & Polruang, S (2022). Life cycle impact assessment and life cycle cost assessment for centralized and decentralized wastewater treatment plants in Thailand. *Scientific Reports, 12*(1), 1–13. https://doi.org/10.1038/s41598-022-18852-y

Pryce, D., Memon, F. A., & Kapelan, Z (2021). Life cycle analysis approach to comparing environmental impacts of alternative materials used in the construction of small wastewater treatment plants. *Environmental Advances, 4*, 100065. https://doi.org/10.1016/j.envadv.2021.100065

Rodríguez, C., Sánchez, R., Rebolledo, N., Schneider, N., Serrano, J., & Leiva, E (2021). Life cycle assessment of greywater treatment systems for water-reuse management in rural areas. *Science of the Total Environment, 795*, 148687. https://doi.org/10.1016/J.SCITOTENV.2021.148687

Romeiko, X (2020). Assessing health impacts of conventional centralized and emerging resource recovery-oriented decentralized water systems. *International Journal of Environmental Research and Public Health, 17*, 1–16. https://doi.org/10.3390/ijerph17030973

Saeed, T., Afrin, R., Muyeed, A., Al, & Sun, G (2012). Treatment of tannery wastewater in a pilot-scale hybrid constructed wetland system in Bangladesh. *Chemosphere, 88*(9), 1065–1073. https://doi.org/10.1016/J.CHEMOSPHERE.2012.04.055

Scholz, M., Morgan, R., & Picher, A. (2007). Stormwater resources development and management in Glasgow: Two case studies. *International Journal of Environmental Studies, 62*(3), 263–282. https://doi.org/10.1080/00207230500034453

Serrano, L., de la Varga, D., Ruiz, I., & Soto, M (2011). Winery wastewater treatment in a hybrid constructed wetland. *Ecological Engineering, 37*(5), 744–753. https://doi.org/10.1016/J.ECOLENG.2010.06.038

Shahabi, M. P., McHugh, A., Anda, M., & Ho, G (2015). Comparative economic and environmental assessments of centralised and decentralised seawater desalination options. *Desalination, 376*, 25–34. https://doi.org/10.1016/j.desal.2015.08.012

Shawkat Mizzouri, N., Ahmed Kochary, S., & Barwari, N (2022). Decentralized sanitation solutions for temporal internally displaced people (IDPs) and refugees camps and residential complexes in Duhok province – Case study. *Ain Shams Engineering Journal, 13*(4), 101706. https://doi.org/10.1016/j.asej.2022.101706

Sheikh, B., Nelson, K. L., Haddad, B., & Thebo, A (2018). Grey water: Agricultural use of reclaimed water in California. *Journal of Contemporary Water Research & Education, 165*(1), 28–41. https://doi.org/10.1111/j.1936-704X.2018.03291.x

Simon Ndibnuh, P. A., David, N. Y., & Ilorme, F (2018). Conceptual design of a DWWT system for the Bamenda Regional Hospital, Cameroon. *Journal of Pollution, 1*(2), 1000110.

Singh, A., Sawant, M., Kamble, S. J., Herlekar, M., Starkl, M., Aymerich, E., & Kazmi, A (2019). Performance evaluation of a DWWT system in India. *Environmental Science and Pollution Research 2019 26:21, 26*(21), 21172–21188. https://doi.org/10.1007/S11356-019-05444-Z

Singh, N. K., & Kazmi, A. A (2018). Performance and cost analysis of DWWT plants in Northern India: Case study. *Journal of Water Resources Planning and Management, 144*(3), 05017024. https://doi.org/10.1061/(ASCE)WR.1943-5452.0000886

Singh, N. K., Kazmi, A. A., & Starkl, M (2015). A review on full-scale DWWT systems: Techno-economical approach. *Water Science and Technology : A Journal of the International Association on Water Pollution Research, 71*(4), 468–478. https://doi.org/10.2166/WST.2014.413

Singh, U., & Colosi, L. (2020). Potable reuse of coalbed methane-produced waters in developing country contexts—Could the benefits outweigh the costs to facilitate coal transitions? *Energies, 13*, 1–18. https://doi.org/10.3390/en13010154

Skambraks, A. K., Kjerstadius, H., Meier, M., Davidsson, Å., Wuttke, M., & Giese, T (2017). Source separation sewage systems as a trend in urban wastewater management: Drivers for the implementation of pilot areas in Northern Europe. *Sustainable Cities and Society, 28*, 287–296. https://doi.org/10.1016/j.scs.2016.09.013

Slompo, N. D. M., & da Silva, G. H. R (2019). Disinfection of anaerobic/aerobic sanitary effluent using ozone: Formaldehyde formation. *Water Environment Research, 91*(9), 898–905. https://doi.org/10.1002/WER.1128

Starkl, M., Brunner, N., Das, S., & Singh, A (2022). Sustainability assessment for wastewater treatment systems in developing countries. *Water, 14*, 1–16. https://doi.org/10.3390/w14020241

SUWANU Europe, 2019. Retrieved April 30, 2021, from https://suwanu-europe.eu/water-reuse-projects-europe/.

Tavarez, J., Dias, M. L., Nunes, I., Duarte, V., Silva, W., & William, H (2020). Compact sewage treatment systems for rural sanitation. *Journal of Urban and Environmental Engineering, 14*(1), 78–86. https://periodicos.ufpb.br/ojs/index.php/juee/article/view/49840/30524

Tchobanoglous, G., Burton, F., & Stensel, H. D (2003). Wastewater engineering: Treatment and reuse. *American Water Works Association Journal, 95*(5), 201.

Tian, H., Wang, X., Lim, E. Y., Lee, J. T., Ee, A. W., Zhang, J., & Tong, Y. W. (2021). Life cycle assessment of food waste to energy and resources: Centralized and decentralized anaerobic digestion with different downstream biogas utilization. *Renewable and Sustainable Energy Reviews, 150*, 111489. https://doi.org/10.1016/j.rser.2021.111489

Tsihrintzis, V. A., Akratos, C. S., Gikas, G. D., Karamouzis, D., & Angelakis, A. N. (2010). Performance and cost comparison of a FWS and a VSF constructed wetland system. *Environmental Technology, 28*(6), 621–628. https://doi.org/10.1080/09593332808618820

USEPA. (2000). *Wastewater technology fact sheet: Oxidation ditches. United States Environmental Protection Agency, Washington, DC.*

Vo, H. N. P., Koottatep, T., Chapagain, S. K., Panuvatvanich, A., Polprasert, C., Nguyen, T. M. H., Chaiwong, C., & Nguyen, N. L. (2019). Removal and monitoring acetaminophen-contaminated hospital wastewater by vertical flow constructed wetland and peroxidase enzymes. *Journal of Environmental Management, 250*, 109526. https://doi.org/10.1016/J.JENVMAN.2019.109526

Vymazal, J. (2019). Constructed wetlands for wastewater treatment. *Encyclopedia of Ecology, 1*, 14–21. https://doi.org/10.1016/B978-0-12-409548-9.11238-2

Wielemaker, R. C., Weijma, J., & Zeeman, G. (2018). Harvest to harvest: Recovering nutrients with new sanitation systems for reuse in urban agriculture. *Resources, Conservation and Recycling, 128*, 426–437. https://doi.org/10.1016/j.resconrec.2016.09.015

Worku, A., Tefera, N., Kloos, H., & Benor, S. (2018). Bioremediation of brewery wastewater using hydroponics planted with vetiver grass in Addis Ababa, Ethiopia. *Bioresources and Bioprocessing, 5*(1), 1–12. https://doi.org/10.1186/S40643-018-0225-5

United Nations World Water Assessment Programme (WWAP). 2017. The United Nations World Water Development Report 2017. Wastewater: The Untapped Resource. Paris, UNESCO.

Zaharia, C. (2017). DWWT systems: Efficiency and its estimated impact against onsite natural water pollution status. A Romanian case study. *Process Safety and Environmental Protection, 108*, 74–88. https://doi.org/10.1016/J.PSEP.2017.02.004

Zang, J., Kumar, M., & Werner, D. (2021). Real-world sustainability analysis of an innovative decentralized water system with rainwater harvesting and wastewater reclamation. *Journal of Environmental Management, 280*, 111639. https://doi.org/10.1016/j.jenvman.2020.111639

9 Feasibility and Suitability of Decentralized Water Systems

Manojkumar Y.[a], *K. Bella*[a], *Venkateswara Rao*[a], *Sridhar Pilli*[a], *and Rajeshwar D. Tyagi*[b,c]

[a] National Institute of Technology, Warangal, India

[b] BOSK Bioproducts, Québec, Canada

[c] Dongguan University, Dongguan, China

9.1 INTRODUCTION

Increasing global water demands, including irrigation, climate change, and population growth, means that existing water resources are under increasing pressure to meet a need that is approaching its capacity limit. Millions of households worldwide rely on centralized water systems. Due to poor maintenance and operation, these systems often fail to reach the poorest and most remote populations. Additionally, the quality and quantity of water provided are often unpredictable. Alternative water supply systems have been tested across a wide range of communities around the globe to determine if they are technologically viable, economically feasible, and socially acceptable (Jones et al., 2006). Sustainability concerns are causing many regions worldwide to look for alternative solutions. In that case, decentralized water systems offer the chance to provide safe drinking water (Moglia et al., 2011). Centralized systems are mostly not feasible and suitable due to some economical, technical, or institutional reasons. Decentralized systems refer to a range of technologies, mostly small-scale water purification and distribution systems. Normally, the supply chain covers individual units to a group of dwellings or a whole community. Also, some systems function as stand-alone systems integrating with the centralized system through satellite systems (Gikas & Tchobanoglous, 2009).

Due to its environmental and financial shortcomings, as well as its lack of resilience, the historical dependence on a linear approach to water infrastructure, imported water, centralized water distribution, energy-intensive treatment, and discharge to surface waters is being called into question. These strategies incorporate source separation, resource recovery, drastically increased water use efficiency, and energy-positive concepts (Garrido-Baserba et al., 2022). In light of rising population demands and supply concerns, water systems need to become more locally robust and sustainable. The technical, environmental, and economic performances of present technologies are frequently thought to be enhanced by decentralized treatment (Garrido-Baserba et al., 2022). The household-scale water systems could result in the emergence of new features, such as the capability of adjusting water quality for specific uses, improved aesthetics, etc.

Decentralized water systems use storm water, roof water, greywater, and local treatment systems. The key benefits of decentralized water systems include less cost, efficient use of natural resources, improved water service security, low risk of failure of water systems, protection of the natural environment, improved community well-being, and local economy strengthening. Moreover, they reduce the reliance on other water sources and reduce the pollutant discharge to the environment, and the availability of other different water servicing choices for the public make them responsive to accomplishing community needs (Sharma et al., 2009). Despite the numerous advantages offered by decentralized systems, several difficulties and uncertainties related to the instalment and operation of such systems prevail. There is a need to establish new rules and regulations for the construction

DOI: 10.1201/9781003376415-11

and use of decentralized water systems. The functional considerations to be laid include requirements for governance development, proposing management and operation models, developing engineering design codes, conducting risk assessments, and a choice selection of different technologies (Sharma et al., 2010).

The use of storm water and greywater harvesting systems through innovative techniques is gaining attention in many countries like India, and China is experiencing inconsistent rainfall and unsustainable consumption by the public due to some populous and hydrologic reasons (Mankad & Tapsuwan, 2011). Decentralized systems are being incorporated into local development standards and building codes as countries begin to legislate water conservation measures (e.g., The National Water Mission established by the Indian Ministry of Jal Shakti) (National Water Mission, 2008). On-site water systems are therefore becoming increasingly important in addressing the global problem of sustainable water use since they provide water to individual homes, small groups of houses, and large-scale housing developments in an economically feasible and simple manner. Due to the immediacy of responsibility in managing decentralized water supplies, the use of these systems is also directly influenced by local attitudes. As most regions affected by water sustainability issues are highly urbanized, it is also imperative that alternative water systems are accepted and adopted by communities in order for their implementation to be successful. Although the developed countries have enough funds to establish new and innovative water supply technologies, there is no assurance that they will get public acceptance. Very little is known about the social and economic factors driving the acceptance of such on-site water treatment and distribution units.

The purpose of this chapter is to analyze in detail the prevailing decentralized water systems, along with understanding the major requirements of establishing a unit. In order to move beyond rhetoric towards an understanding of the capacity requirements of decentralized systems, it is necessary to review previous implementation successes and failures and to identify the factors that contribute to superior performance. In addition, further validation of the system must be conducted in order to ensure that its performance is in accordance with its design and is appropriately applied in the social-technical context.

9.2 DECENTRALIZED WATER SYSTEMS

A decentralized water supply involves purifying and distributing water on a small scale. They offer the possibility to provide reliable and safe drinking water to informal and rural settlements (Piratla & Goverdhanam, 2015). Even if the main water supply chain is centralized, the decentralized units can supply water for other purposes like agricultural use, vehicle washing, street washing, etc. The major sources of decentralized water systems are surface water sources like lakes; rivers, which are the most frequently used ones when the water supply is at the communal level; groundwater sources like drilled wells; and recycled water like wastewater or rainwater, collected through a recycling system (Chirisa et al., 2017). The advantages and disadvantages of a decentralized water system are shown in Figure 9.1.

The decentralized water systems can be household water treatment systems (HWTS) or small-scale treatment systems (SSTS). HWTS supplies water after filtering either through a small unit close to taps or through a larger storage tank. HWTS are ideal for the sustainable use of recycled storm water and wastewater. HWTS are of two types classified based on the quantity of water supplied: point of use (POU) and point of entry supply (POE). The POU supply systems supply at least 25 L/d for each household. All water entering a household is treated by a point of entry system, which is typically an additional purification process. SSTS collect, treat, and supply water locally. This can be done at a smaller level (neighbourhood, community, or commercial level). The SSTS systems supply 1000–10,000 L/d for communities. They are often supplied to emergency camps (SSWM, 2020).

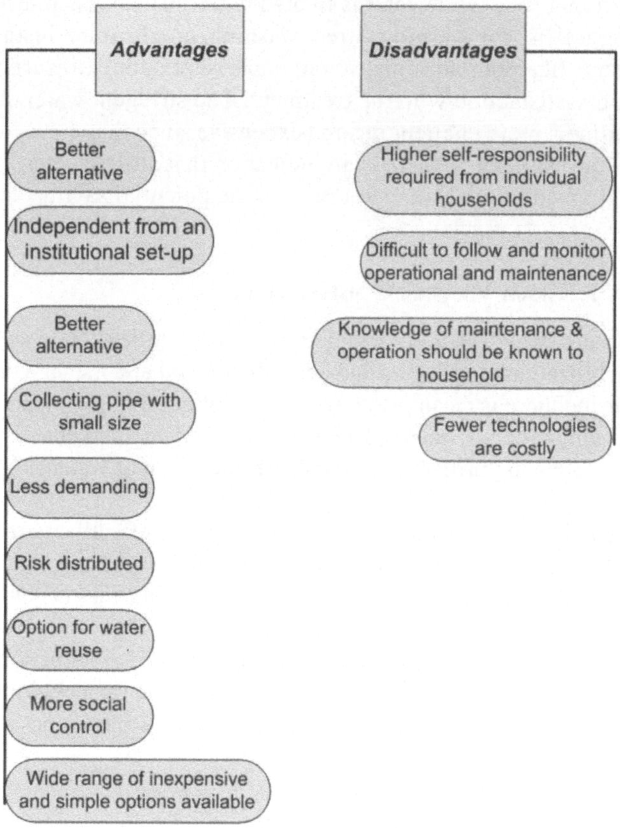

FIGURE 9.1 Advantages and disadvantages of the decentralized system.

9.3 CASE STUDIES OF SUCCESSFUL DECENTRALIZED SYSTEMS AROUND THE WORLD

In many developing countries, unplanned reuse technologies for untreated wastewater in agriculture exist. Water-scarce regions of the world have also been reusing treated wastewater at a significant scale for three decades. Most of these systems are decentralized systems whose output is enjoyed by downstream farmers (Lazarova & Asano, 2013).

9.3.1 Case Study 1: Decentralized Wastewater Treatment System in Bengaluru

The city of Bengaluru, the capital of Karnataka state in India, has an established decentralized wastewater treatment system and reuse (DWRTU) where 100% wastewater treatment, as well as reuse, happens, thereby implementing zero-waste liquid discharge (ZWLD) (Evans et al., 2014). As per Karnataka State Pollution Control Board (KSPCB), around 2200 residual complexes within the city implemented the system on their premises (KSPCB, 2012). The city is pumping almost 1250 MLD of water from the Cauvery River, located 100 km away and 300 m downhill. The groundwater source of supply covers around 600 MLD. The city started facing heavy water shortages due to the sudden fall of groundwater levels, and with withdrawals from the Cauvery River, the entire water supply was allocated for domestic water reuse (DMG, 2011). After the ZWLD order issued by KSPCB in 2004, a stringent water treatment and reuse facility was established in 2000 buildings by 2014, with a treatment capacity of 110 MLD (Evans et al.,

2014). The major portion of the wastewater is treated to tertiary levels, and 70% of treated water is reused for flushing toilets, car washing, street washing, etc. In some instances, decentralized treatment had problems like poor administration, unpleasant odours, insufficient attention, STP failure, and, finally, dissatisfaction with the treatment. The stringent water quality requirements set out make the treatment more challenging and expensive since they were overly strict for some parameters. Despite the challenges, there is an indication that attitudes may be shifting as locals see the advantages of an additional water source and the potential savings on their water bills.

9.3.2 Case Study 2: Gram Vikas in Odisha, India

The Gram Vikas system, which is a civil society organization developed in the Odisha state of India, enables the mobilization of villagers and helps them build and manage their own piped water supply systems, ensuring there is clean water available continuously (Chowdhury & Santos, 2010). Integrated water and sanitation facilities reduce women's drudgery and enable safe sanitation practices. They follow the following principles in the decentralization of water resources.

- The inclusive participation of the community ensures the effective participation of every household, including the poorest and socially excluded.
- The entire community shares money for the operation and maintenance of the system, hence making the community easily accountable.
- Convergence is generated, local contributions are mobilized, construction is managed, and operations and maintenance are handled by people from the very beginning.
- In-built financial stability through the formation of the surplus fund as well as maintenance fund.

9.3.3 Case Study 3: Decentralized Wastewater System in Harare, Zimbabwe

Decentralized wastewater management systems are present in Zimbabwe, mostly in rural regions where pit latrines are often utilized. It is normal practice in areas like Gweru, Redcliff, Nemanwa, and Mupandawana to handle wastewater by decentralized treatment. The decentralized wastewater system in Gweru began operation in 1994. A typical wet technology utilized in places with big plot sizes is septic tanks. Aquaculture, man-made wetlands, and duckweed pond systems are all employed at Redcliff. Decentralized wastewater management systems, septic tanks, pit latrines, and urine diversion are some of the solutions that are becoming more and more common. Less-concentrated wastewater can also be treated naturally using ponds with algae and duckweed as well as artificial wetlands that also gather protein-rich biomass. Other methods for stabilizing organic materials include the use of maturation ponds and anaerobic pre-treatment. In Harare, decentralized systems also incorporate algae-based technologies. This comprises four treatment lines of tiny waste stabilization ponds made of algae that are dispersed across the Mabvuku/Tafara high-density area. Several limitations had to be considered before a decentralized wastewater management plan could be adopted in Harare. First, Harare already had a substantial infrastructure that was functioning and represented a sizable capital commitment. Therefore, a progressive approach was needed, and the existing system could not be phased out immediately. Few residential areas make use of shared toilets and water amenities. Second, Harare's land use patterns have so many distinct characteristics that it would be unreasonable to apply a single solution to all of them. Residential communities with low, medium, and high densities have very distinct lifestyles and water consumption patterns. Therefore, a differentiated strategy was needed. The social, financial, and environmental benefits of decentralization became crucial considerations when considering this type of scheme in developing countries' urban water management plans, such as for Harare, particularly in the peri-urban areas, informal settlements, and southern suburbs where wastewater collection and/or treatment is not available (Chirisa et al., 2017).

9.3.4 Case Study 4: Decentralized Wastewater Treatment Plants in Bangkok, Thailand

Decentralized treatment scenarios have lower environmental costs than centralized treatment scenarios. Eight centralized WWTPs (Bangsue, Chatuchak, Chongnonsi, Dindaeng, Nongkaem, Rattanakosin, Sipraya, and Thungkru plants) and 12 decentralized WWTPs (Bangbua, Bangna, Bonkai, Huai-khwang, Hua-mark, KhlongChan, KhlongToei, RamIntra, RomKlao, ThaSai, and Tungsonghong I and II) are now present in Bangkok, the capital of Thailand. Dewatering is used to remediate the sludge in the decentralized WWTPs for dry organic waste. Sludge is delivered by truck to Nongkeam WWTP for centralized wastewater treatment after thickening, where it is converted into biogas and fertilizer. Sludge is sun-dried for decentralized wastewater treatment and then utilized for landfill. Therefore, composting should be used to remediate sludge. Construction of a centralized WWTP takes two years and costs 3358.27 million THB; a decentralized WWTP takes one year and costs 118.95 million THB, according to the Department of Drainage and Sewerage of the Japan International Cooperation Agency. Compared to decentralized treatment, the average input of centralized treatment is greater and includes transportation, water supply, and energy. The centralized treatment generates more sludge than the decentralized treatment, which accounts for the greater levels of heavy metals in those cases. The centralized treatment scenarios have higher overall costs than the decentralized treatment scenarios.

9.4 SYSTEM REQUIREMENTS FOR DECENTRALIZED WATER SYSTEMS

For the successful implementation of a decentralized water system, the following conditions are considered to be suitable:

- Where clusters comprising on-site systems exist and the fate of pollutants has not been cared for.
- Localities or communities where water scarcity prevails.
- Localities where the reuse of treated water can be done at the local level.
- Areas where a centralized system cannot be extended.
- Newly developed industrial parks, residential flats, etc.

Further requirements are discussed in Table 9.1.

TABLE 9.1
Basic Requirements of a Decentralized Water System

S. No.	Impact	Details
1	Health and hygiene	The human risk of exposure to pathogens should be reduced.
2	Functionality	Easy operation, monitoring, and functionality maximization.
3	Natural and environmental resources	The resources available for construction and operations and resulting emissions must be monitored.
4	Financial and economic	The system should run as per the capital availability of villagers.
5	Sociocultural and institutional	Social acceptance, perception, convenience, and compliance with legal networks must be considered.
6	Aesthetics and nuisance	Since the units are located near residential areas, the nuisance level should be minimum.

9.5 SUSTAINABILITY INFLUENCING FACTORS FOR THE TREATMENT SYSTEMS

The two main elements of a sustainable decentralized water system are natural-technical and managerial-institutional. The natural-technical element includes the requirement for greater investment in source sustainability, appropriate technology support for demand management, and weather information to help farmers choose their crops more efficiently. The information about weather conditions helps the villagers to plan the system more efficiently. The decentralization of technical skills is also crucial. The managerial element covers the empowerment of the gram panchayats as the governing institutions. If there is a convergence of resources at the panchayat level, then government investments will contribute to the system's security.

Decentralized water recycling and reuse might be developed as a long-term solution to supply people in these areas with safe, palatable, and inexpensive water, particularly in developing nations where there is a physical and economic water shortage. The availability of technology, the location of the implementation, and the potential for expansion are just a few of the numerous aspects that must be taken into account when choosing an acceptable system for implementation (Maniam et al., 2022). Installation and operational costs, ease of deployment, maintenance, acceptability, comfort, performance, energy requirements, supply chain requirements, environmental effect, odour, and health considerations are the elements that are typically viewed to influence the choice of technology.

9.6 ECONOMIC EVALUATION OF DECENTRALIZED TREATMENT, RECLAMATION, AND REUSE

The idea of environmental accounting is applied to the field of economic evaluation in order to accurately assess the economic effectiveness of the decentralized systems and represent the economic gains. The economic analysis takes into account the projects' primary economic, environmental, and social implications. Decentralized wastewater reuse systems are economically viable, according to the studies. This implies that the systems have a beneficial impact on society. Therefore, it makes sense for the government or community to develop decentralized wastewater reuse systems (Liang & van Dijk, 2010).

The feasibility of implementing new water supply systems is heavily influenced by economic efficiency. The economic analysis of decentralized water systems depends on various factors like the type of system, source of water, consumer number scale, water policies, the intended use of water, etc. Generally, there are two types of costs and benefits associated with decentralized water systems: the direct costs associated with the infrastructure and the indirect costs associated with the environmental externalities associated with its implementation. Regulation and delivery of more water provides direct benefits to the region, as well as indirect benefits associated with better environmental flows (Arora et al., 2015). The economical evaluation of these systems is greatly influenced by special and temporal scales. The special scale ranges from individual dwellings to a cluster of communities or housing colonies. The temporal scale can be evaluated through an LCA which assesses all the events related to the financial implications of the decentralized water system. Depreciation and improvements to a system can be included in these assessments.

Several studies have been carried out by researchers analyzing the economic viability of decentralized water systems. In lieu of expanding existing centralized water infrastructure or trucking water, Rabaey et al. (2020) evaluated the feasibility of installing a household-scale water system based on rainwater capture and greywater reuse. The authors used a membrane bioreactor system followed by an ultraviolet disinfection unit for greywater treatment. The treated water is passed through a reverse osmosis unit in order to reuse treated water as drinking water. The system had an initial payback of ten years. Also, the household system could generate new pathways for use of water for other purposes like softened water for cleaning, gardening, laundry, etc.

Cost-benefit analyses of wastewater treatment and reuse projects have to be conducted with a focus on decentralized systems, which are attracting significant interest worldwide, particularly in water-scarce nations and areas. The cost-benefit analysis of the decentralized water reclamation and reuse project in a recently constructed residential neighbourhood in Xi'an, China, was conducted using the net benefit value model. When all of the treated water is reused for non-potable uses like gardening and artificial pond replenishment, the system with the dual-pipe collection and greywater treatment and reuse was determined to be economically unsuitable. The project's economic advantage becomes increasingly considerable when further environmental advantages are taken into account (Chen & Wang, 2009).

The cost of implementing on-site water harvesting and recycling systems would raise the overall construction costs of multi-family buildings by roughly 6% and single-family homes by about 12%, according to a thorough economic analysis that took CapEx and OpEx into account. The estimated total cost of on-site water provision (including harvesting, treatment, recycling, and monitoring) for buildings or combined water systems with more than 300 occupants ranged from \$1.5/m^3 to \$2.7/m^3, which is significantly less than the typical tariffs collected by utilities in the United States and Western Europe. Water expenses might be significantly lower in locations where buildings can avoid having to link to centralized sources of potable water and sewage disposal (Garrido-Baserba et al., 2022). The most cost-effective decentralized solutions may be found for structures with populations greater than 300 people, such as big buildings, neighbourhoods, or city blocks. However, district-scale techniques that are decentralized and use fewer structures (leading to lower population densities) could not produce the same economic benefits (Garrido-Baserba et al., 2022). The costs to developers and citizens will be equivalent to or lower than those of present systems, according to a thorough techno-economic and feasibility review of the costs of extreme decentralization, which will also result in a decrease in per capita water demand, utilizing current on-site treatment methods and technology, which are anticipated to become mature in the next two decades (Garrido-Baserba et al., 2022).

Although economically viable, Beijing's decentralized wastewater reuse systems are not financially viable. It was discovered that a significant factor in the system's infeasibility on a financial level is the low fee actually charged for recycled water. If the financial issues are not resolved, Beijing's decentralized wastewater reuse systems could cease to function. It is now theoretically possible to create water of any quality because of advancements in wastewater reuse technology. Small wastewater reuse systems may now produce recovered water in a trustworthy manner (Liang & van Dijk, 2010). The overall economic analysis of decentralized systems includes the analysis of both financial and economic costs and benefits.

9.7 OVERALL EVALUATION OF THE DECENTRALIZED TECHNOLOGIES AND THEIR SUCCESS

Decentralized solutions provide safer and more affordable water reuse choices for communities without centralized wastewater treatment facilities. Decentralized systems have investment, operating, and maintenance costs that are 20% to 50% or even cheaper than centralized systems. Thus, potential small-scale methods for water reuse, including rainwater and greywater recycling, are more feasible. The practicality and durability of creating decentralized systems, particularly in developing nations, might be impacted by several variables. Decentralized water reuse system deployment is particularly difficult in nations with low water rates (Maniam et al., 2022).

The financial and environmental effects of decentralized, energy-efficient wastewater treatment systems are evaluated. A power law function was used to modify the investment and operating cost functions that were developed. The corresponding investment costs for the various wastewater systems servicing villages with between 50 and 250 inhabitants were from 400 to 200 euros per person and from 70 to 20 euros per person per year, respectively. To compare the effects of various wastewater treatment technologies on the environment, a life cycle analysis method was employed. The evaluation concentrated on two conventional and two energy-saving technologies (constructed

TABLE 9.2
Cost-benefit Analysis of Decentralized Treatment Systems

S. No.	Parameter	Cost	Benefits
1	Environmental	Air pollution Nuisance	Water conservation
2	Economic	Capital cost Operation cost Maintenance cost	Reduction of distribution, transportation, and treatment costs Pollutants reuse
3	Social	Health hazards	Water availability and awareness

wetlands and slow rate infiltration) (activated sludge process). Energy-efficient wastewater treatment systems have little influence on the environment, making them the most pertinent global warming indicators. Materials used in system construction and system operating lifespan were evaluated as options for reducing life cycle effects. Therefore, the total environmental effect of such design choices shouldn't be underestimated given the enormous potential of energy-saving wastewater treatment systems (Nogueira et al., 2009).

The reclamation and reuse of municipal wastewater encounters many regional transversal difficulties including societal perception and a dearth of effective models for business and government. From a purely technological and economic perspective, it is essential to guarantee the water quality requirements set out by regulations and end users, and to be able to show effective fit-for-use water reclamation trains. Water reuse is still far from its potential, despite its obvious advantages. According to the economic analysis, using municipal reclaimed water is more cost-effective than using tap water, saving the waste management sector between 0.13 and 0.52 euros per cubic metre and the electrocoating business between 0.49 and 0.98 euros per cubic metre (Echevarría et al. 2022). Beyond the economic outcomes, additional factors need to be taken into account to ensure water supply, such as the preservation of freshwater resources and climate change adaptation methods (Echevarría et al. 2022). The overall costs and benefits of decentralized treatment systems are shown in Table 9.2.

9.8 FUTURE IMPLEMENTATION STRATEGIES

The future of wastewater reuse depends on the public acceptability of decentralized systems, and negative public opinion might threaten upcoming wastewater reuse initiatives. A thorough assessment of the community's technological and self-sufficiency capabilities must go hand in hand with the choice of any decentralized treatment technique.

In order to achieve sustainability objectives or to fulfil the community's future water supply demands, decentralized wastewater treatment and reuse may be a good solution. The new management framework advocated places a strong emphasis on decentralization, but putting it into practice will be difficult due to issues with planning, design, and operation, as well as policy and community involvement. Determining which kind and how to include decentralized choices in an existing portfolio of water supply is often a difficult decision-making process. Fortunately, a lot of water utilities are including supply portfolios that include reclamation and reuse.

When choosing the best sanitation technique for a specific circumstance, innovations in decentralized systems might be quite helpful. Decentralized treatment facilities could be used in places without sufficient wastewater treatment facilities or toilets in order to totally address the issue of open defecation and sanitation since they are not only effective but also affordable, which is a key factor for developing nations (ArorA et al., 2016).

Though selecting the "most appropriate technology" is a difficult process, doing so might lower the likelihood of future issues and failures. The two most important criteria for selecting a treatment technique are cost and suitability (Massoud et al., 2009).

Integrating decentralized reuse can be a potent way to achieve dependability and sustainability beyond water supply diversification, whether thinking about new systems or the extension of current systems. The ability to take decentralized water reuse into account for the next water resource management planning can be improved by research. Because of its advantages on the social and environmental fronts, decentralized systems are becoming a more practical and perhaps even desirable addition to the current centralized infrastructure.

It is also necessary to analyze the techniques or procedures used to choose the decentralized system, including community involvement or educational initiatives, its governance structure (public/private), and any potential advantages or hazards to the community.

9.9 CONCLUSIONS AND PERSPECTIVES

Existing technologies may be applied to develop commercially viable systems that significantly decrease the requirement for centralized utilities and, in certain cases, do eliminate the need for centralized sewage or water delivery. Decentralized treatment solutions, which can include sewer mining, on-site non-potable or potable reuse systems, artificial wetlands, and several other alternatives, are being used since they are less expensive than centralized reuse and distribution systems. Decentralized systems for collecting rainfall and treating greywater for reuse are anticipated to play a significant part in reducing reliance on centralized water supplies and supplying water for urban agriculture and even informal settlements. Alternative decentralized strategies that could increase the economic and environmental sustainability, dependability, and resilience of urban water systems have to be developed. Decentralized wastewater treatment systems may, in some circumstances, be practical and even advantageous. The effectiveness and long-term survival of these systems depend on careful design, site evaluation, technology selection, and community engagement. The other factors that decide the feasibility of decentralized systems are population density, location, site conditions, technology selection, operation and maintenance, regulations, cost considerations, environmental impact, public acceptance, scale of treatment system, and climate and weather conditions. Decentralized treatment systems are highly suitable for remote and rural areas, small- and medium-sized communities, site constraint places, conservation of water, reduced infrastructure cost, reduced energy consumption and cost, quick deployment, innovative technologies, and reduced environmental impact. They can offer practical and cost-effective wastewater treatment solutions, particularly in places where centralized methods are impractical or too expensive. A decentralized wastewater treatment system's applicability depends on a thorough site evaluation, consideration of local laws, support from the community, selection of the right technology, and continuing maintenance. Decentralized systems can offer effective and sustainable wastewater treatment solutions for diverse scenarios with the correct planning and execution. The social and economic aspects influencing the acceptability of such on-site water treatment and distribution systems are poorly studied, which might hinder the promotion of decentralized treatment systems.

REFERENCES

Arora, Jitesh, Anuj Yadav, and Devendra Saroj. "Potential of decentralised wastewater treatment systems applicable to India." *Current World Environment* 11, no. 2 (2016): 338.

Arora, Meenakshi, Hector Malano, Brian Davidson, Rebecca Nelson, and Biju George. "Interactions between centralized and decentralized water systems in urban context: A review." *Wiley Interdisciplinary Reviews: Water* 2, no. 6 (2015): 623–634.

Chen, R., and X. C. Wang. "Cost–benefit evaluation of a decentralized water system for wastewater reuse and environmental protection." *Water Science and Technology* 59, no. 8 (2009): 1515–1522.

Chirisa, Innocent, Elmond Bandauko, Abraham Matamanda, and Gladys Mandisvika. "Decentralized domestic wastewater systems in developing countries: The case study of Harare (Zimbabwe)." *Applied Water Science* 7, no. 3 (2017): 1069–1078.

Chowdhury, Imran, and Filipe Santos. "Scaling social innovations: The case of Gram Vikas." In *Scaling social impact*, pp. 147–166. Palgrave Macmillan, New York, 2010.

DMG. "Groundwater Hydrology and Groundwater Quality in and around Bangalore City." *Technical Report* (2011).

Echevarría, C., M. Pastur, C. Valderrama, J. L. Cortina, A. Vega, C. Mesa, and M. Aceves. "Techno-economic assessment of decentralized polishing schemes for municipal water reclamation and reuse in the industrial sector in costal semiarid regions: The case of Barcelona (Spain)." *Science of the Total Environment* 815 (2022): 152842.

Evans, Alexandra EV, Samyuktha Varma, and A. Krishnamurthy. "Formal approaches to wastewater reuse in Bangalore, India." (2014).

Garrido-Baserba, Manel, Irene Barnosell, Maria Molinos-Senante, David L Sedlak, Korneel Rabaey, Oliver Schraa, Marta Verdaguer, Diego Rosso, and Manel Poch. "The third route: A techno-economic evaluation of extreme water and wastewater decentralization." *Water Research* 218 (2022): 118408.

Gikas, Petros, and George Tchobanoglous. "The role of satellite and decentralized strategies in water resources management." *Journal of Environmental Management* 90, no. 1 (2009): 144–152.

KSPCB. (2012). A report of monitoring committee on performance of wastewater treatment plants in the apartments and other establishments in BBMP limits of Bangalore.

Lazarova, Valentina, and Takashi Asano. *Milestones in water reuse*. IWA publishing, 2013.

Liang, Xiao, and Meine Pieter van Dijk. "Financial and economic feasibility of decentralized wastewater reuse systems in Beijing." *Water Science and Technology* 61, no. 8 (2010): 1965–1973.

Maniam, Geetha, Nur Ain Zakaria, Choe Peng Leo, Vasilena Vassilev, Karen Banahene Blay, Kourosh Behzadian, and Phaik Eong Poh. "An assessment of technological development and applications of decentralized water reuse: A critical review and conceptual framework." *Wiley Interdisciplinary Reviews: Water* 9, no. 3 (2022): e1588.

Mankad, Aditi, and Sorada Tapsuwan. "Review of socio-economic drivers of community acceptance and adoption of decentralised water systems." *Journal of Environmental Management* 92, no. 3 (2011): 380–391.

Massoud, M. A., A. Tarhini, and J. A Nasr. (2009). Decentralized approaches to wastewater treatment and management: Applicability in developing countries. *Journal of Environmental Management, 90*(1), 652–659.

Moglia, Magnus, K. S. Alexander, and Ashok Sharma. "Discussion of the enabling environments for decentralised water systems." *Water Science and Technology* 63, no. 10 (2011): 2331–2339.

National Water Mission, 2008. Comprehensive mission document. Volume – II. https://nwm.gov.in/sites/default/files/nwm28756944786.pdf.

Nogueira, R., A. G. Brito, A. Patrícia Machado, P. Janknecht, J. J. Salas, L. Vera, and G. Martel. "Economic and environmental assessment of small and decentralized wastewater treatment systems." *Desalination and Water Treatment* 4, no. 1-3 (2009): 16–21.

Piratla, Kalyan R, and Suraj Goverdhanam. "Decentralized water systems for sustainable and reliable supply." *Procedia Engineering* 118 (2015): 720–726.

Rabaey, Korneel, Tom Vandekerckhove, Arjen Van de Walle, and David L Sedlak. "The third route: Using extreme decentralization to create resilient urban water systems." *Water Research* 185 (2020): 116276.

Sharma, Ashok K, Andrew L Grant, Tim Grant, Francis Pamminger, and Lisa Opray. "Environmental and economic assessment of urban water services for a greenfield development." *Environmental Engineering Science* 26, no. 5 (2009): 921–934.

Sharma, Ashok, Stewart Burn, Ted Gardner, and Alan Gregory. "Role of decentralised systems in the transition of urban water systems." *Water Science and Technology: Water Supply* 10, no. 4 (2010): 577–583.

Sustainable Sanitation and Water Management (SSWM). 2020. Decentralised supply. Decentralised Supply | SSWM - Find tools for sustainable sanitation and water management!

10 Decentralized Wastewater Treatment and Reuse toward Sustainability and Circular Economy

Bhoomika Yadav[a], *Shraddha Chavan*[a], *Rajeshwar D. Tyagi*[b,c], *Patrick Drogui*[a], *and Jonathan W.C. Wong*[b]

[a] Université du Québec, Québec, Canada

[b] Dongguan University of Technology, Dongguan, China

[c] BOSK Bioproducts, Québec, Canada

10.1 INTRODUCTION

The increasing levels of contamination and frequency of shortages of water have necessitated the need for sustainability in water resources management. One of the greatest challenges in the water and wastewater management sector is the implementation of economic treatment facilities for the human settlements in rural, urban, and peri-urban areas that will permit the reuse of treated effluents in various agricultural and industrial applications, thus reducing the overall user demand for water resources (Jhansi & Mishra, 2013). Decentralized wastewater treatment systems in accordance with the local government are being recognized as a suitable way to contribute towards the areas with no access to clean water supply or basic sanitation as well as increasing the efficiency of wastewater treatment and treated water recovery and reuse. The treatment technologies should be efficient, reliable, low cost, low maintenance, and locally acceptable (Libralato et al., 2012).

Wastewater is now considered as a renewable resource which can be used for potable and non-potable applications (agricultural irrigation, toilet flushing, industrial processes, ground water recharge etc.) as well as to derive energy and other value-added products. Increasing the accessibility of water and sanitation does not mean overexploitation of resources but improving the water management and treatment system by reducing, recycling, and reusing the materials, resources, and the products as well as identifying new water sources such as storm water, greywater, and reclaimed wastewater for potential reuse and recycle (Libralato et al., 2012). Decentralized systems are more flexible and are easily adaptable to local conditions of rural areas or low-income areas of cities where centralized systems are not functional/applied.

In some countries, the small, decentralized plants can treat greater volume of wastewater than existing centralized treatment plants. In Venice, Italy, there are more than 140 small decentralized biological wastewater treatment systems as well as various septic tanks (Tromellini, 2008).

This chapter discusses several options to achieve sustainability in wastewater treatment and management. It highlights the main aspect of decentralized wastewater treatment systems with respect to building a circular economy and sustainability in the environment. Low-cost, on-site decentralized systems have been discussed with the approach for zero-waste discharge strategy to build circularity in the process. Moreover, waste-to-energy strategies and treated wastewater reuse

DOI: 10.1201/9781003376415-12

have been described in detail. Further, the challenges to implement these strategies for wastewater reuse for sustainable water management have also been underlined.

10.2 MOVING TOWARDS CIRCULAR ECONOMY USING DECENTRALIZED SANITATION

Decentralized wastewater systems are usually implemented for smaller areas such as individual houses, clusters of homes, or district-level settings. They use different water sources, such as rainwater or greywater, and are focused on non-potable purposes such as toilet flushing, irrigation, etc. (Sgroi et al., 2018). Unlike centralized systems, decentralized greywater reuse, wastewater recycling, and rainwater harvesting allow the treatment at the source, which reduces the costs associated with collection and transportation. Installation of reuse systems of rainwater, greywater, and hybrids have been performed in countries like Brazil, Germany, and Australia to capture and reuse non-potable water (Ghafourian et al., 2022). Although centralized systems have been widely used in the past decades, research on sustainability and cost-benefit analyses suggest shifting to either hybrid or fully decentralized systems to build sustainability and reliability in wastewater management (López-Serrano et al., 2023). Decentralized systems are regarded as more sustainable as compared to centralized systems because they emphasize water conservation and reuse, improve the resilience of water networks and infrastructure, and reduce costs on repair and maintenance, leading to environmental and local economic development (Sgroi et al., 2018).

One of the widely used decentralized technologies is nature-based systems such as constructed wetlands (CWs), which, apart from meeting the strict physical and chemical regulations of water treatment and reuse, require low energy, are cost effective, have low maintenance, and are easy to operate (López-Serrano et al., 2023). Researchers have also shown that the greenhouse gas (GHG) emissions of CWs are 2714 times lower than traditional wastewater treatment plants (WWTPs) (Liu et al., 2019). CWs link their high efficiency with the ecological benefits that can improve water supply from a sustainable point of view.

Studies have shown by installing rainwater harvesting and greywater reuse systems in two residential houses (A & B) in Brazil, the potable water saving efficiency could reach 35.5% and 33.6% for rainwater harvesting and 30.4% and 25.6% for greywater reuse for house A and B, respectively. Whereas, on using both these systems together, the water saving efficiency reached to 36.4% and 33.8% for house A and B, respectively (Ghisi & de Oliveira, 2007). In another study, the water performance of putative hybrid rain-greywater system in commercial and residential structure in Malaysia was estimated (Leong et al., 2017). The results showed that such hybrid systems in a commercial building should predominantly use both greywater and rainwater to fulfil the water demand, whereas hybrid rain-greywater systems in residential settings should prioritize rainwater harvesting and reuse.

The principles of circular economy focus on the closed loop concept where resources, energy, and materials are used multiple times by lowering, reprocessing, recycling, and retrieving materials in production and consumption phases (Yadav et al., 2021). Researchers have shown various decentralized technologies to reclaim water and capture intrinsic values in wastewater such as nutrients, energy, materials, etc., that can be used in various industrial, agricultural, and urban applications (Sgroi et al., 2018). However, new policies based on circular economy principles are needed for a paradigm shift to establish a better and cost-effective sustainable model in wastewater management.

10.3 LOW-COST, LOW-MAINTENANCE ON-SITE SYSTEMS

There are several options available for on-site wastewater treatment, among which on-site anaerobic treatment has been widely employed due to its simplicity, cost-effective reactor design, small space requirement, low sludge production, low energy and nutrient demand, and energy production.

Moussavi et al. (2010) studied on-site anaerobic treatment of medium-strength residential wastewater in a pilot-scale up-flow septic tank. The effect of hydraulic retention times on up-flow septic tank performance was studied. Overall, the system was concluded to be technically and economically promising to serve as an alternative to conventional septic tanks for on-site decentralized treatment of residential wastewater, especially in rural settings.

Disposal of poorly treated wastewater from on-site systems that serve households and small communities can cause pollution of the environment, causing harmful risks to humans. It is important to routinely detect and quantify the levels of pathogens in the on-site sanitation systems designed to contain and treat the household wastewater. Amin et al. (2020) in their study detected high concentrations of pathogens being discharged from the commonly used on-site sanitation systems (septic tanks and anaerobic baffled reactors) and their transport to the nearby drainage or waterways. Therefore, improved management and maintenance, further treatment of the effluent, and strict monitoring of regulations are needed to protect public health. Researchers have upgraded the performance of septic tanks for on-site sewage treatment with cost-effectiveness (Jorsaraei et al., 2014). The quality of treated wastewater reached the standard level of irrigation by changing the structure of a conventional septic tank and making it an advanced septic reactor. This was done by modifying the flow direction from horizontal to up-flow and applying distribution trays.

The critical situation of domestic wastewater treatment in rural settlements of Brazil requires development of on-site systems using cheap and sustainable materials that can be applied for long-term small and isolated rural communities. Researchers have investigated a full-scale on-site domestic wastewater treatment system composed of septic tank, an anaerobic filter with green coconut husk, and an intermittent sand filter (de Oliveira Cruz et al., 2019). The effluent generated by this system was in accordance with the regulations and could be reused for agricultural activities. Also, using coconut husks in anaerobic filters can provide new and more sustainable use of this material, which is generally treated as a solid waste.

On-site wastewater treatment and recycling systems are sustainable, as the wastewater is treated near the source and can be reused. A large amount of research is now being carried out on on-site wastewater treatment. Studies have found that most of the systems suffer inefficiency in operation and maintenance after the responsibility is passed to the local community. Studies should be carried out for determining the best configurations of decentralized and on-site treatment facilities and operating at low cost with sustainable wastewater treatment systems which meet the regulations. Future research should focus on techno-economic assessment of field installation of decentralized systems to find appropriate technology for a particular area and compare the new technologies with those already existing.

In many developing countries, centralized wastewater systems cover a portion of urban areas and are not implemented in smaller towns and densely populated low-income areas of the cities. On-site systems are not appropriate in slum areas or denser settlements. Decentralized wastewater treatment systems connected to sewer systems or communal sanitation stations can close the gap between on-site and centralized systems. Decentralized systems managed by the community can help in swift sanitation improvements in high priority areas where local government cannot provide full sanitation or treatment services (Eales et al., 2013). In Indonesia, thousands of decentralized wastewater treatment systems have been developed that are managed by community. It was also indicated that the cost of community sanitation centres with decentralized wastewater treatment was 50% lower than the centralized treatment system (Eales et al., 2013).

10.4 ZERO-WASTE DISCHARGE SYSTEMS

Zero-waste discharge is a holistic waste management notion that recognizes waste as a resource and a symbol of inefficiency in the society. In the zero-waste discharge strategy, after the end-of-life phase of the product, the zero-waste product could either be reused, repaired, or remanufactured to produce secondary product. Or the zero-waste discharge systems ensures that the discarded waste could either be recycled, recovered, or reused without polluting the environment, thus following sustainability by optimizing the use of natural resources with minimum environment degradation (Zaman, 2014).

Researchers adopted a net-zero approach to solve the water stress challenge of Mumbai for water treatment and its reuse (Raut, 2019). They concluded that sequential batch reactor to be most sustainable technology due to its less-energy requirements, easy operation, adaptability to wastewater, and high treatment efficiency to achieve net positive water management on-site. Net-zero waste strategy can be developed by creating opportunities to increase water efficiency by creating a closed loop in the system.

Nansubuga et al. (2016) proposed a decentralized wastewater treatment system with zero waste generation. They used an M and M (major and minor) sewage treatment concept by separating wastewater near to the source into two distinct streams: a major liquid stream consisting of 90% of the flow and a minor solid stream with 10% of the flow. Pre-concentration of the sludge can be performed by a high-rate oxidation sludge system. After separation, the major flow stream is treated with trickling filters, which can be made from charcoal that is locally available to release a final effluent that can be reused for crop irrigation, park irrigation, and other non-potable uses. The minor flow stream that contains solids can be anaerobically digested with the sludge from the major flow stream to recover biogas that can be further used by locals for energy (cooking, lighting, electrical energy for the wastewater plant). The residue digestate can be sun-dried and used as a fertilizer. The charcoal used for cleaning the major wastewater stream can be later dried and used for its original purpose. This system achieves a closed loop, recycling the available resources that would lead to sustainable treatment of the wastewater.

10.5 WASTE-TO-ENERGY STRATEGIES

Given that wastewater contains nutrients, and energy, it is also known as "resource water", with a high possibility of energy production, storage, and recovery primarily in three forms: organic carbon (chemical (bio)energy recovery through the production of biogas), heat (recovery of thermal energy from wastewater), and ground water as a sustainable energy source. In existing techniques, the power potential is partially extracted from wastewater (Verstraete et al., 2009). Typically used activated sludge treatment facilities use a considerable amount of power for aeration. The chemical energy in the form of chemical oxygen demand (COD) is converted into metabolic heat during oxidation. The sludge is generally digested to retrieve some of the energy. The Netherlands produced 95 million Nm^3 (or a theoretical value of 2215 TJ) of biogas in 2006 (El-Fadel & Massoud, 2001). In a combined power and heating system, this biogas is transformed into heat and energy. Relative to WWTP without sludge digestion, for instance, pre-settling and sludge digestion, 40% less net energy is used. With anaerobic treatment techniques, more energy may be retrieved. However, the fundamental issue with direct anaerobic treatment is the excessive dilution of the sewage (Fernandes et al., 2010).

Large quantities of energy are introduced to wastewater via heating in households for bathing, cleaning, cooking, etc. The need for hot water heat increases as we develop additional high-performing structures and accounts for a sizable portion of the energy consumption. Currently, residences are being constructed to use only 750 m^3 of natural gas annually; as a result, warm water will account for around 50% of domestic gas use. The sewer systems still receive heated water discharges, making household wastewater a heat transporter. About 60% of the drinkable water in Dutch homes is heated to a predetermined temperature (Nederlof & Frijns, 2010).

For instance, the temperature of the water used for showers is 38°C, while the temperature of the water used for laundry is 40–60°C (Blokker et al., 2013). The mean temperature of the wastewater of the residence, which varies throughout the day, is 27°C (van Schaik et al., 2010). Because there are 7.2 million houses in the Netherlands, the overall heat capacity of treated sewage wastewater is 56,000 TJ/y, with the greatest heat flux capacity in wastewater from Dutch households being 21.3 MJ/home day. The electricity use of the water industrial sector would be equal to retrieving 25% of this heat.

Meggers and Leibundgut (2011) established a novel concept for the quick recovery of heat from warm water inside homes. The system comprises a heat recovery tank that receives hot wastewater discharged from appliances like showers/baths and washing machines. The heat pump receives its heat from a heat exchanger, which raises the temperature of the recovered heat to a level high enough to produce fresh hot water. The prospect of directly replenishing the hot water supply through wastewater heat recovery arises as a result of this. The idea came from studies on low-energy building systems.

Centralized heat recovery along sewer lines or WWTPs is simpler to set up than heat recovery strategies inside buildings. Heat exchangers can be built in the sewer to recover thermal energy and use it to heat surrounding neighbourhoods or commercial buildings. This has been put into reality at the housing estate level in Hamburg and Zurich. Lower wastewater temperatures as a result of the heat recovery from the sewage system, however, could have a negative impact on the effectiveness of wastewater treatment operations, particularly nitrification processes (Wanner et al., 2005). The expected temperature effect might be minimized, since dissolving oxygen requires less energy at lower temperatures. It is necessary to conduct more research on heat recovery and the impact of temperature on WWTPs.

10.6 RECYCLING WATER FOR NON-POTABLE USES

Inadequate water quality regulations have frequently made it difficult to reuse wastewater. The rules for urban uses of recycled water often only include the water-quality requirements needed for greywater. However, the current legal framework of these regulations differs from one nation to another (Behling et al., 1997). The effects of decentralization on wastewater management schemes go beyond the potential adoption of effective technology for local resource recycling and recovery and the improvement of ecosystem sustainability. Strategy, decision-making, the design of infrastructure networks, and procedures for system operation and maintenance are all influenced by the selection of decentralization (Behrends et al., 2007). Realistically speaking, this could help to increase the influence of residents on municipal authorities, solve health and ecological issues that are specific to the area, and result in better services and more responsive service providers. Smaller, decentralized systems are more likely to be used to implement innovation in alternate technology and practices if it becomes essential, as capital investment for decentralized systems is typically lower than for centralized ones (Capodaglio, 2017).

Decentralized systems require integral and essential modifications in the way plans and decisions about the management of water resources, as well as infrastructure, operations, and maintenance, are made, even though they are intended to have positive effects on the management of urban water systems (Capodaglio et al., 2016). Due to conventional, training, and habit, resistance to this fresh approach is, therefore, to be expected (even from technical professionals). Decentralized systems will typically be in compliance with existing water use (reuse) needs, in which domestically treated water and its nutrient content could boost agricultural output or (in more urban areas) be used as a replacement for treated water supply for ground water resource resupply (US Environmental Protection Agency, Office of Water Management, 1997). Reuse possibilities for locally reclaimed water are listed in Figure 10.1. Based on the local circumstances and conditions, some of these possibilities might only be available during certain seasons. By maintaining the wastewater management component to a minimum and concentrating mostly on treating and disposing of the minimized amounts of wastewater, decentralization emphasizes a holistic strategy, maximizing the advantages of reducing the amount of waste at source and enhancing reprocessing or reuse at the site (Istenic et al., 2015). Especially in tiny areas with low population densities, it has been predicted that collection and treatment costs can be lowered by more than 60% in comparison to those of a centralized system (Lew et al., 2011).

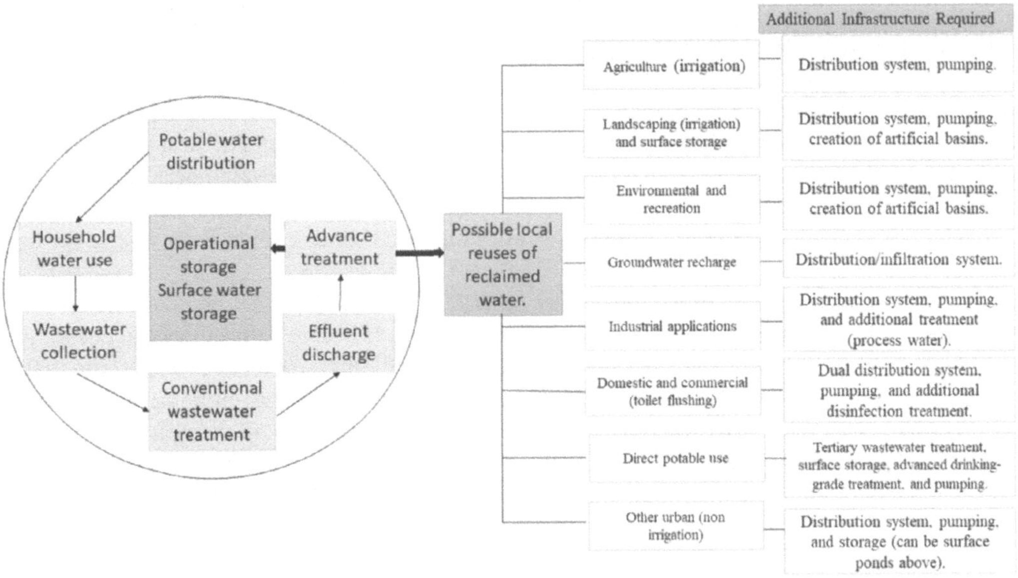

FIGURE 10.1 Available local reuses of recycled water.

10.7 THE ULTIMATE GOAL OF SUSTAINABILITY

The idea of sustainability seeks to preserve economic prosperity, save the environment, manage natural resources wisely, and advance social equality. It also entails the necessity to develop systems that will prevent future deterioration of life quality due to the utilization of natural assets (Imperatives, 1987). Table 10.1 illustrates the primary concern factors that need to be considered when assessing the technological viability of wastewater treatment methods (as well as the relationships between various subcategories). Using a set of specially created sustainability indicators, Muga and Mihelcic (2008) compared the sustainability assessment of mechanical, lagoon, and land treatment methods for wastewater treatment on case studies meant for communities with a wastewater flow of less than 20,000 m^3/day. The findings demonstrated that the choice and perception of markers are impacted by the regional and demographic characteristics of the site, and the total viability of wastewater treatment technology is a measure of financial, ecological, and social aspects.

Segregation of various household wastewater streams with a goal of diversified on/off-site treatment for resource recovery/recycling is recognized as the most interesting idea to re-establish sustainable management of carbon (Otterpohl et al., 2002). The main argument in favour of source separation is that concentrated streams make it simpler and more effective to extract necessary nutrients and power, which are typically found in blackwater. Source separation benefits could be further increased by using intensive water-saving techniques, such as low-flashtubes created expressly to lower the amount of water used for the mobilization of waste sediments (Eriksson et al., 2002).

The combination of source separation and extremely low-flush devices (like those found in airplanes) are presently being established for residential use, and non-traditional transportation techniques (i.e., vacuum sewers) (Kujawa-Roeleveld & Zeeman, 2006). This would have a significant impact on the present conceptual frameworks of wastewater disposal. Although vacuum sewage systems (VSS) have been in use for a while, their potential as a sustainable long-term technical solution for wastewater collection has only lately come to light (Capodaglio, 2017). The primary goal of urban sewerage was to prevent human contact with faeces or any other contaminated water. Vacuum sewers can accomplish this goal while providing complete environmental and health protection, minimizing seepage and odours, and lowering energy use and operational expenses (Panfil et al., 2013).

TABLE 10.1
Challenges Concerning Sustainability with the Majority of Decentralized Wastewater Treatment Systems

Decentralized Wastewater Treatment	Health Factor	Environmental Factor	Technological Factor	Financial Factor	Socio-cultural Factor
Constructed Wetlands Systems	Possibly configured for solar disinfection (post-treatment)	Modified naturalistic system Nearly neutral in terms of energy Works well with less populated areas Minimal resource extraction (but potential harvesting of plants)	Very simple to use High resilience and minimal crisis susceptibility High expandability, if possible, adaptability High water loss during warmer weather	Investing costs were primarily for the land plot	Adoption is positive if it is "out of the way" and doesn't cause trouble Potentially inadequate institutional comprehension (nonstandard behaviour)
Aerobic Conventional	Need post-treatment	Standard technology Potential for tertiary recovery of energy and nutrients from sludge	Remotely controlled operation is reasonably simple Moderate resilience and vulnerability (toxicity from discharges, power outages) Expansion is conceivable at moderately high expense Appropriate for smaller facilities' more affordable "package" construction	Costly operations and maintenance (energy and sludge management)	Acceptance based on circumstances and previous performance Potential odour annoyance Reputable institution
Membrane Bioreactor Aerobic	Possibility of reusing without further processing	Energy-consuming Lower footprint than traditional aerobics More effectiveness Possibility of tertiary nutrient recover	More difficult operation, with timely fouling issues Strong against changes in flow and load, moderately susceptible to power outages, and less poisonous Substantial expenditures required for progress Ideal for small sites' more affordable "package" construction	Highest O&M expenditures (more energy, but less sludge to handle)	Acceptance based on circumstances and previous performance Potential odour annoyance Accepted \with expense concerns institutionally

The idea behind VSS is to receive wastewater and transmit it to the last station, where it can either be dispatched for local treatment or subsequently transported to a centralized or decentralized WWTP. This is done by employing the differential pressure in a vacuum-maintained pipeline (Terryn et al., 2014). In VSS, a vacuum is created at a singular, centralized location, requiring just one source of energy usage. This eliminates the requirement for inspection manholes, simplifies energy sources, and lowers construction expenses and regular operating costs. In agricultural villages in Egypt, a study examined precise hydrological layouts for traditional and VSS, assessing the cost elements of both processes using data study (Little, 2004). The study concluded that the capital cost of vacuum systems was consistently found to be cheaper than that of gravity systems, even though the overall prices of gravity and VSS systems both depend on the population and area covered. The study suggested the construction of vacuum sewerage based solely on the observation of the resulting environmental benefits, even though the analysis of operating and maintenance costs was not fully completed (Ma et al., 2015).

10.8 CHALLENGES IN THE PROCESS

According to the US Environmental Protection Agency (EPA), decentralized wastewater facilities "may serve as a cost-effective and long-term solution for satisfying human safety and water quality standards, especially in less densely populated regions". Planning, design of infrastructure networks, and procedures for system operation and maintenance are all influenced by the choice of decentralization (Crites et al., 2010). Practically speaking, it can help to increase the influence of residents in local government and address environmental and health issues that are specific to the area, resulting in better services and more responsive service providers (Arias et al., 2020). Decentralized systems will typically be in line with regional water use (reuse) needs, where domestically treated water and its nutrient availability could boost crop yields or (in even more urban areas) serve as an alternative for treated supply water for landscaping or sustainable ground water resources (Frank & Rhynus, 1920).

Several non-technological obstacles stand in the way of wider adoption of decentralized wastewater treatment systems in industrialized countries, in addition to issues like land availability (particularly for constructed wetlands), pricing, and environmental requirements that can be resolved by selecting a more appropriate decentralized technology. Consequently, decentralized wastewater treatment systems are more widely accepted in emerging nations where there is no prior technology development, favourable economic circumstances, and the influence of younger, more innovative engineers who favour that option (Alvarado et al., 2017). The principal barriers to more widespread adoption of decentralized systems in industrialized nations are primarily pre-existing centralized settings and conventional methodologies preferred by senior public-system water technologists (Capodaglio, 2017). The proponents of decentralized wastewater treatment technologies are primarily modern young professionals who struggle to persuade decision-makers and conventional wastewater specialists with a "business as usual" approach to adopting such technologies. Decentralized systems must (1) safeguard human health and the environment; (2) dispel the stigma attached to on-site systems' (Cashman et al., 2018) efficiency; (3) contribute significantly to sustainable water management; (4) support paradigm shift from effluent disposal to water reuse; and (5) be merged with current centralized systems in order to overcome the demands of the 21st century.

There are numerous chances for merging decentralized treatment systems with current centralized systems as the urban and suburban economy expands. There is a range of potential outcomes that could alter the function of current facilities, lead to more effective treatment procedures, and improve understanding of the value of water (Daigger, 2009). Detailed functional facility reusing with solids processing at centralized treatment programs, on-site septic tanks with effluent pumping to centralized facilities (STEP systems), and remote monitoring and administration of decentralized systems at centralized facilities are a few instances of integrated systems. Through the correct application of current technologies, integrated decentralized treatment systems will deliver suitable quality water for a specific process (Eggimann et al., 2016).

10.9 CONCLUSIONS AND PERSPECTIVES

There is an urgent need for a more sustainable society, with new developments to efficiently handle the goals for recovering all the resources available in wastewater. Sustainable wastewater management can promote recovery of high-value resources such as energy, fertilizer, and water by integrating resource recovery technologies in wastewater treatment systems. Decentralized wastewater treatment systems can not only reduce the effects of wastewater disposal on the environment in small rural settlements but also increase the reuse of wastewater. To complement the decentralized systems, it is important to consider whether to introduce source separation system or other water-saving systems, which can be site specific for enhancing resources and energy recovery. New policies based on a circular economy will cause paradigm shifts starting with the principle of segregation at the source to build a sustainable model in wastewater treatment and management with enhanced resource recovery. The application of the decentralized system approach to wastewater treatment and reuse requires planning, design, and operation in a holistic manner, considering all the subsystems and their impacts on the environment to ensure circularity and sustainability in terms of economic and environmental aspects.

Currently, there is a good knowledge regarding implementation and performance of decentralized systems; however, technological transfer is still lacking. Moreover, inappropriate choice of technology, incorrect design technology, and lack of proper maintenance can bring negative health and environmental impacts. The research about the barriers to using decentralized treatment to support circular economy implementation in the water sector is still limited. The barriers to the decentralized system have not been identified in detail through quantified studies. Also, each region has its own economic, environmental, and social aspects, so the identified problems may not be applied throughout. When both centralized and decentralized systems are viable, both the systems should be researched in various parameters to find the most appropriate and site-specific system that is economical, sustainable, and locally acceptable.

REFERENCES

Alvarado, A., Larriva, J., Sánchez, E., Idrovo, D., Cisneros, J.F. 2017. Assessment of decentralized wastewater treatment systems in the rural area of Cuenca, Ecuador. *Water Practice and Technology*, **12**(1), 240–249.

Amin, N., Liu, P., Foster, T., Rahman, M., Miah, M.R., Ahmed, G.B., Kabir, M., Raj, S., Moe, C.L., Willetts, J. 2020. Pathogen flows from on-site sanitation systems in low-income urban neighborhoods, Dhaka: A quantitative environmental assessment. *International Journal of Hygiene and Environmental Health*, **230**, 113619.

Arias, A., Rama, M., González-García, S., Feijoo, G., Moreira, M.T. 2020. Environmental analysis of servicing centralised and decentralised wastewater treatment for population living in neighbourhoods. *Journal of Water Process Engineering*, **37**, 101469.

Behling, E., Diaz, A., Colina, G., Herrera, M., Gutierrez, E., Chacin, E., Fernandez, N., Forster, C. 1997. Domestic wastewater treatment using a UASB reactor. *Bioresource Technology*, **61**(3), 239–245.

Behrends, L., Bailey, E., Jansen, P., Houke, L., Smith, S. 2007. Integrated constructed wetland systems: Design, operation, and performance of low-cost decentralized wastewater treatment systems. *Water Science and Technology*, **55**(7), 155–161.

Blokker, E., van Osch, A., Hogeveen, R., Mudde, C. 2013. Thermal energy from drinking water and cost benefit analysis for an entire city. *Journal of Water and Climate Change*, **4**(1), 11–16.

Capodaglio, A.G. 2017. Integrated, decentralized wastewater management for resource recovery in rural and peri-urban areas. *Resources*, **6**(2), 22.

Capodaglio, A.G., Ghilardi, P., Boguniewicz-Zablocka, J. 2016. New paradigms in urban water management for conservation and sustainability. *Water Practice and Technology*, **11**(1), 176–186.

Cashman, S., Ma, X., Mosley, J., Garland, J., Crone, B., Xue, X. 2018. Energy and greenhouse gas life cycle assessment and cost analysis of aerobic and anaerobic membrane bioreactor systems: Influence of scale, population density, climate, and methane recovery. *Bioresource Technology*, **254**, 56–66.

Crites, R.W., Middlebrooks, E.J., Reed, S.C. 2010. *Natural wastewater treatment systems*. CRC press.

Daigger, G.T. 2009. Evolving urban water and residuals management paradigms: Water reclamation and reuse, decentralization, and resource recovery. *Water Environment Research*, **81**(8), 809–823.

de Oliveira Cruz, L.M., Gomes, B.G.L.A., Tonetti, A.L., Figueiredo, I.C.S. 2019. Using coconut husks in a full-scale decentralized wastewater treatment system: The influence of an anaerobic filter on maintenance and operational conditions of a sand filter. *Ecological Engineering*, **127**, 454–459.

Eales, K., Blackett, I., Siregar, R., Febriani, E. 2013. Review of community-managed decentralized wastewater treatment systems in Indonesia. Water and Sanitation Program Technical Paper. Washington DC: World Bank; 2013. https://openknowledge.worldbank.org/entities/publication/0ba328cb-e784-5c2b-a6a3-30de4938801a

Eggimann, S., Truffer, B., Maurer, M. 2016. The cost of hybrid waste water systems: A systematic framework for specifying minimum cost-connection rates. *Water Research*, **103**, 472–484.

El-Fadel, M., Massoud, M. 2001. Methane emissions from wastewater management. *Environmental Pollution*, **114**(2), 177–185.

Eriksson, E., Auffarth, K., Henze, M., Ledin, A. 2002. Characteristics of grey wastewater. *Urban Water*, **4**(1), 85–104.

Fernandes, B.S., Peixoto, G., Albrecht, F.R., del Aguila, N.K.S., Zaiat, M. 2010. Potential to produce biohydrogen from various wastewaters. *Energy for Sustainable Development*, **14**(2), 143–148.

Frank, L.C., Rhynus, C.P.. 1920. The Treatment of Sewage from Single Houses and Small Communities, Public Health Bulletin No. 101, United States Public Health Service, Washington, DC.

Ghafourian, M., Nika, C.-E., Mousavi, A., Mino, E., Al-Salehi, M., Katsou, E. 2022. Economic impact assessment indicators of circular economy in a decentralised circular water system—Case of eco-touristic facility. *Science of the Total Environment*, **822**, 153602.

Ghisi, E., de Oliveira, S.M. 2007. Potential for potable water savings by combining the use of rainwater and greywater in houses in southern Brazil. *Building and Environment*, **42**(4), 1731–1742.

Imperatives, S. 1987. Report of the World Commission on Environment and Development: Our common future. *Accessed Feb*, **10**, 1–300.

Istenic, D., Bodík, I., Bulc, T. 2015. Status of decentralised wastewater treatment systems and barriers for implementation of nature-based systems in central and eastern Europe. *Environmental Science and Pollution Research*, **22**(17), 12879–12884.

Jhansi, S.C., Mishra, S.K. 2013. Wastewater treatment and reuse: Sustainability options. *Consilience* **10**(1), 1–15.

Jorsaraei, A., Gougol, M., Van Lier, J.B. 2014. A cost effective method for decentralized sewage treatment. *Process Safety and Environmental Protection*, **92**(6), 815–821.

Kujawa-Roeleveld, K., Zeeman, G. 2006. Anaerobic treatment in decentralised and source-separation-based sanitation concepts. *Reviews in Environmental Science and Bio/Technology*, **5**(1), 115–139.

Leong, J.Y.C., Oh, K.S., Poh, P.E., Chong, M.N. 2017. Prospects of hybrid rainwater-greywater decentralised system for water recycling and reuse: A review. *Journal of Cleaner Production*, **142**, 3014–3027.

Lew, B., Lustig, I., Beliavski, M., Tarre, S., Green, M. 2011. An integrated UASB-sludge digester system for raw domestic wastewater treatment in temperate climates. *Bioresource Technology*, **102**(7), 4921–4924.

Libralato, G., Ghirardini, A.V., Avezzù, F. 2012. To centralise or to decentralise: An overview of the most recent trends in wastewater treatment management. *Journal of Environmental Management*, **94**(1), 61–68.

Little, C. 2004. A comparison of sewer reticulation system design standards gravity, vacuum and small bore sewers. *Water SA*, **30**(5), 137–144.

Liu, D., Zou, C., Xu, M. 2019. Environmental, ecological, and economic benefits of biofuel production using a constructed wetland: A case study in China. *International Journal of Environmental Research and Public Health*, **16**(5), 827.

López-Serrano, M.J., Lakho, F.H., Van Hulle, S.W., Batlles-delaFuente, A. 2023. Life cycle cost assessment and economic analysis of a decentralized wastewater treatment to achieve water sustainability within the framework of circular economy. *Oeconomia Copernicana*, **14**(1), 103–133.

Ma, X., Xue, X., González-Mejía, A., Garland, J., Cashdollar, J. 2015. Sustainable water systems for the city of tomorrow—A conceptual framework. *Sustainability*, **7**(9), 12071–12105.

Meggers, F., Leibundgut, H. 2011. The potential of wastewater heat and exergy: Decentralized high-temperature recovery with a heat pump. *Energy and Buildings*, **43**(4), 879–886.

Moussavi, G., Kazembeigi, F., Farzadkia, M. 2010. Performance of a pilot scale up-flow septic tank for on-site decentralized treatment of residential wastewater. *Process Safety and Environmental Protection*, **88**(1), 47–52.

Muga, H.E., Mihelcic, J.R. 2008. Sustainability of wastewater treatment technologies. *Journal of Environmental Management*, **88**(3), 437–447.

Nansubuga, I., Banadda, N., Verstraete, W., Rabaey, K. 2016. A review of sustainable sanitation systems in Africa. *Reviews in Environmental Science and bio/technology*, **15**, 465–478.

Nederlof, M., Frijns, J. 2010. Zero impact water use in the built environment. *Towards 0-impact buildings and built environments*. Techne Press, Amsterdam, 199–208.

Otterpohl, R., Braun, U., Oldenburg, M. 2002. Innovative technologies for decentralised wastewater management in urban and peri-urban areas. *Berichte-Wassergute Und Abfallwirtschaft Technische Universitat Munchen Berichtsheft*, **173**, 109–126.

Panfil, C., Mirel, I., Szigyarto, I., Isacu, M. 2013. Technical, economical, social and ecological characteristics of vacuum sewage system. *Environmental Engineering & Management Journal (EEMJ)*, **12**(5).

Raut, N.V. 2019. Assessment of decentralized wastewater treatment systems on net zero approach for high-density residential buildings in Mumbai, India. *Proceedings of the International Conference of Architectural Science Association*. pp. 363–372.

Sgroi, M., Vagliasindi, F.G., Roccaro, P. 2018. Feasibility, sustainability and circular economy concepts in water reuse. *Current Opinion in Environmental Science & Health*, **2**, 20–25.

Terryn, I.C.C., Lazar, I., Nedeff, V., Lazar, G. 2014. Conventional vs. vacuum sewerage system in rural areas—An economic and environmental approach. *Environmental Engineering and Management Journal*, **13**(8), 1847–1859.

Tromellini, E. 2008. Centralised control for Venice's decentralised waste-water treatment. *Water*, **21**.

US Environmental Protection Agency, Office of Water Management. 1997. *Response to Congress on use of decentralized wastewater treatment systems*. US Environmental Protection Agency, Office of Water.

van Schaik, A., Reuter, M.A., van Stokkom, H., Jonk, J., Witter, V. 2010. Management of the web of water and web of materials. *Minerals Engineering*, **23**(3), 157–174.

Verstraete, W., Van de Caveye, P., Diamantis, V. 2009. Maximum use of resources present in domestic "used water." *Bioresource Technology*, **100**(23), 5537–5545.

Wanner, O., Panagiotidis, V., Clavadetscher, P., Siegrist, H. 2005. Effect of heat recovery from raw wastewater on nitrification and nitrogen removal in activated sludge plants. *Water Research*, **39**(19), 4725–4734.

Yadav, B., Atmakuri, A., Chavan, S., Tyagi, R.D., Drogui, P., Pilli, S. 2021. Role of Bioeconomy in Circular Economy. in: *Biomass, Biofuels, Biochemicals*, Elsevier, pp. 163–195.

Zaman, A.U. 2014. Identification of key assessment indicators of the zero waste management systems. *Ecological Indicators*, **36**, 682–693.

11 Challenges and Future Perspectives of Decentralized Water Treatment

Bhukya Gopal, Chinthala Sumanth, Sridhar Pilli, and Manojkumar Y.
Water and Environment Division, Department of Civil Engineering, NIT, Warangal, Telangana, India

11.1 INTRODUCTION

Wastewater management is an essential practice for present and future generations to meet the increasing water demand. The consumption of water is increasing with increases in population, urbanization, and industrialization. The quantity of wastewater generated from these activities is also increasing. Efficient wastewater management is necessary to prevent the water pollution that causes adverse effects on human health and the environment. On the other hand, rural wastewater management systems also needed to be improved to avoid untreated discharges into water bodies. Water pollution, due to discharge of wastewater in rivers, lakes, and ponds causes biological contamination, which is responsible for waterborne diseases and is harmful to aquatic life. Current wastewater management is mostly using centralized treatment units. Centralized treatment facilities are provided for treating bulk quantities of wastewater from cities, towns, and industries. The drawbacks of centralized wastewater treatment are (a) land requirement, (b) continuous monitoring, (c) skilled workers are required, (d) cost of operation and maintenance, (e) energy-consuming process, (f) sludge management, (g) approximately 65–75% cost of the project is spent on sewers to transport the wastewater from source of generation to treatment facility, (h) flexibility in terms of redesign and expansion for future load is not possible, and (i) loss due to major or minor accidents is huge. In order to minimize the land cost, transportation of wastewater, maintenance of sewers, and energy consumption in the centralized systems, decentralized wastewater treatment is an emerging and efficient alternative solution for centralized treatment facilities (British Columbia Small Water Systems 2022; Chirisa et al. 2017).

11.2 ADVANTAGES OF DECENTRALIZED CONCEPT

Decentralized wastewater management can be defined as "collection, treatment and reuse of wastewater from individual houses, clusters of homes, communities, industries and individual commercial facilities at or near the point of waste generation" (Tchobanoglous et al. 2004). Decentralized water treatment can be considered as green, sustainable, and environmentally friendly. We can achieve zero discharge goals by implementing decentralized concepts. Figure 11.1 shows the decentralized water treatment at or near the point of waste generation.

Some of the advantages of a decentralized system are as follows:

1. Easily applicable in complex terrain. Generally, cities are growing bigger and bigger, and as a result, the terrain of the cities may not be uniform. Under these circumstances, transporting waste or water will be energy intensive. Instead, if a decentralized facility is installed for smaller clusters that have similar terrains, the transport of waste in the nearby cluster can be carried out using gravity, and the treated waste can be put into use locally at a small scale.

DOI: 10.1201/9781003376415-13

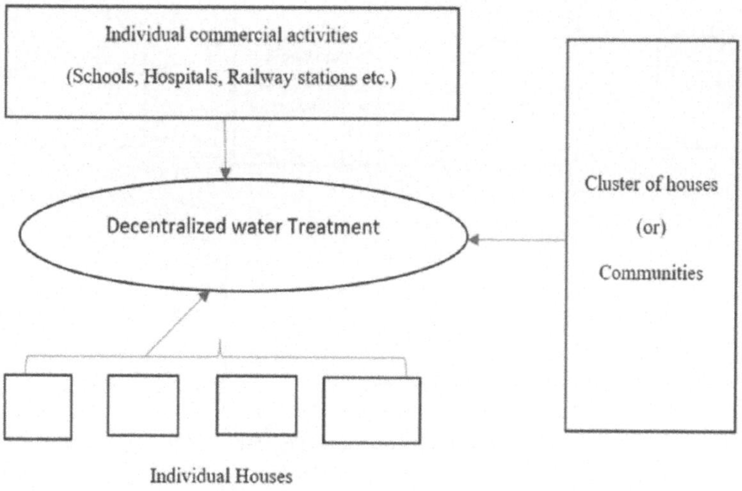

FIGURE 11.1 Decentralized water treatment facility at point of waste generation.

2. Integrating the systems to solve other urban challenges. In many instances, municipalities need treated wastewater for applications including sprinkling at nearby construction sites, high-traffic road intersections, air pollution hot spots, cleaning of few local areas, etc. Under such circumstances, transporting water from a large centralized system will not be economical. A decentralized system integrated with the purpose of water usage in the city can effectively solve this problem in all dimensions.
3. Improves treatment of waste. Decentralized treatment units give the decision-makers a lot of opportunities to manage the waste at the local level. For instance, if industrial and domestic sewage is mixed and sent to a centralized treatment unit, the products of the treatment (both solids and liquids) may contain many impurities and make the products unsuitable for any application. Decentralized treatment facilities provide opportunities to reuse the end products generated from the treatment units and improve the chances to treat and reuse the waste at the local level.
4. Other advantages. Decentralized wastewater treatment can be installed at or near the source of waste generation, which minimizes or eliminates the transportation of wastewater through sewers. The land requirement is less compared to centralized facilities, and the assigned land can be utilized further due to its underground design. It can be designed on gravity to minimize the energy consumption by eliminating pumps and pipes. Continuous monitoring and skilled workers are not required. The treated water can be reused for gardening and flushing purpose. Water demand can be reduced. It provides flexibility for new treatment units. Quality of treatment is high due to less and similar type of waste. Water pollution in rivers, lakes, and ponds can be reduced due to treated discharge. The effect of pollution on ambient air quality and ecosystem can be minimized. Ground water contamination and effect on environment due to wastewater is significantly minimized (AquaTechSystems 2021; USEPA 2022).

11.3 CONTINUED DEVELOPMENTS IN THE DECENTRALIZED APPROACH

Decentralized wastewater systems are designed to reuse treated water for on-site applications. In general, the treatment units used in decentralized wastewater treatment are septic tanks, anaerobic baffled reactors, constructed wetlands, and filtration. A typical flow diagram of decentralized wastewater treatment is shown in Figure 11.2.

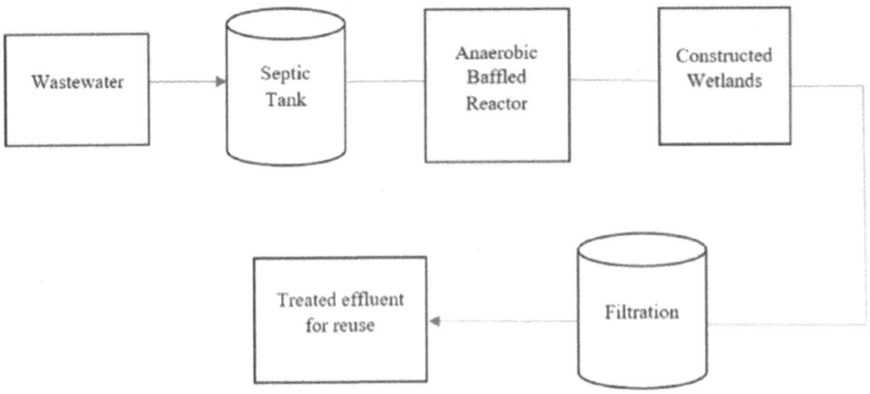

FIGURE 11.2 Decentralized water treatment process flow diagram.

In recent times, various combinations of treatment units have been studied and implemented under decentralized wastewater treatment. They include phytorid technology, soil biotechnology, tiger biofilters, fixed film biofilters, waste stabilization ponds, up-flow anaerobic sludge blankets, moving bed biofilm reactors, and trans biofilters. Considering the factors such as availability of land, pollution load, discharge standards, and application of treated water, an effective decentralized treatment facility can be implemented by using the preceding treatment units. In many cases, after primary treatment, constructed wetlands are used, which can provide oxidation in the treatment process and can be further used as a feedstock for animals. Many countries that are focusing on efficient wastewater management are exploring the potential of decentralized wastewater treatment. Various combinations of treatment units are studied and implemented in various applications across India (CSE INDIA 2022). In countries with limited water resources and increasing water demand, decentralized wastewater treatment can be an effective option to reuse water for various applications.

11.4 MAJOR GAPS IN THE SYSTEM

11.4.1 Limitations in Existing Technologies

Many existing technologies are developed based only on a particular growth rate of population. For many reasons, the population of the cities started expanding many-fold. The population prediction was nowhere close to the current value due to various socio-economic factors. As a result, the treatment units are unable to cater to the needs of the rising population, and the demand for more treatment units is constantly increasing.

Second, many emerging pollutants have added to the list and the norms have become more stringent. Under these situations, the existing units need to be retrofitted with additional treatment systems. In many cases, the availability of land is minimum, and retrofitting the existing technologies with new units has become more and more challenging.

With the increase in the pollutant load and other factors, many treatment units have reached closer to the end of their lives and they need a lot of maintenance. Lack of proper expertise is resulting in the reduced efficiency of the treatment units.

Due to the location of the existing treatment units, many newly built areas were forced to connect with the existing units to maintain effluent norms. In some cases, sewage pumping has become mandatory and it has significantly increased the cost of operation.

11.4.2 Need for New Infrastructure Solutions

Since availability of space is a big challenge in urban areas, we need more infrastructure solutions that can treat the waste using minimum space.

Additionally, the existing treatment units may see their end of life in the coming years. As they get close to the end, the operation and maintenance costs increase, resulting in an economic burden. In many instances, the equipment's spare parts will no longer be available. For instance, we may use certain kinds of pumps in our treatment unit, and if there is a problem with any pump and the company no longer manufactures the necessary parts, the entire pump may have to be replaced. Hence new infrastructure solutions are the need of the hour.

11.4.3 Lack of Implementation of Policies and Regulations

Every city must always keep an eye on the following:

1. The impact due to any Government Order (GO) passed by the central government or state government
2. The impact due to any court judgment given by any court
3. The impact due to the reduction in any budgetary allocations for any reasons
4. The impact due to the removal of any government servants or any related manpower in the operations
5. The impact due to the change in any of the terms of the contract associated with the operation of the waste treatment units

Due to a lack of coordination between various agencies, lack of enough resources, and for many other reasons, many policies and regulations are not carried out.

11.4.4 Integration with Existing Centralized Systems

Many cities have already constructed wastewater treatment plants in the periphery of the cities, depending up on the population, land availability, and other factors. To integrate the decentralized units with the existing centralized systems, feasibility studies have to be carried out to understand the following:

1. Which sector/area of the city is consuming more energy for the treatment of the waste
2. Which sector/area of the city has enough land to accommodate more decentralized systems
3. In which direction is the city expanding and where are new units needed
4. From which point can the sewage be diverted using gravity to the nearest decentralized system to reduce the load on the centralized unit
5. What fraction of the waste can be diverted to the nearby new decentralized unit with minimum cost
6. How to optimize the transport of the treated effluent to various nearest commercial or residential units to minimize the operation costs

11.4.5 Other Challenges

The decentralized treatment facility should meet the sewage or wastewater discharge standards, and needs to adapt to the evolving discharge standards. Innovation in the design of treatment units and filter media are required for high efficiency optimization of decentralized systems for high organic loading rate and hydraulic flows, efficient sludge management, and resource recovery. Integration with centralized systems is required for effective sludge management. Systems should provide consistent quality with less energy consumption. Life cycle assessment of the project allows establishment of an economical and environmentally friendly treatment facility by studying parameters like energy, cost, and emissions of the project. Further, it requires effective policy-making and awareness about the technology. The policy should encourage the individual and communities towards the implementation of technology. Effective policy is essential

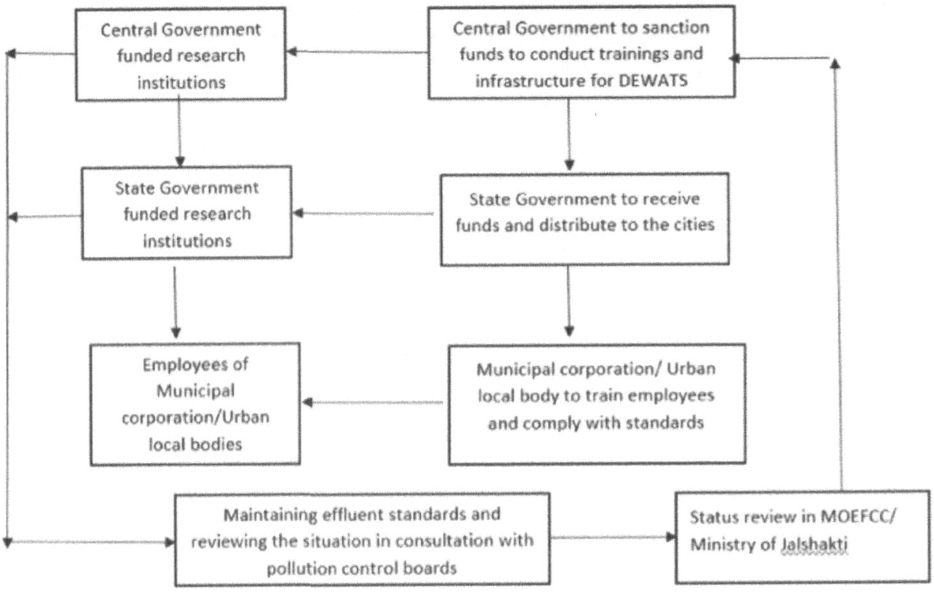

FIGURE 11.3 Implementation of decentralized treatment systems: A possible structure.

for monitoring system performance and discharge standards. In order to promote the technology at a rural scale, respective authorities should step forward to provide incentives and awareness programs for implementing decentralized wastewater treatment systems considering their economic status. There is a need to clear the misconception among the communities regarding odour issues and health effects. Government monitoring and management is important for effective application (Starkl et al. 2018). The possible pathway to implement the decentralized systems has been shown in Figure 11.3.

11.5 HOW TO FURTHER OPTIMIZE THE INTEGRATED DECENTRALIZED WATER TREATMENT SYSTEMS

The following guidelines may be implemented to optimize the integrated decentralized treatment systems:

1. Include the construction of decentralized systems in the master plan that is being developed for the cities. The master plan should allocate some space for construction of decentralized treatment systems so that the implementation is possible in the coming years. The possible methodology for implementing the same is shown in Figure 11.4.
2. Identification of product consumers like truck terminals, parks, etc., which consume water and mapping them with the nearest decentralized unit may reduce the operation costs significantly and optimize the overall system.
3. The flushing system of public toilets provided in each and every commercial building need reused water for flushing to reduce the overall consumption of water for cities. A local pipeline network for a smaller domain to supply reused water to flushing systems may reduce the water consumption of the city in the long run.
4. The city is already divided into various units, including commercial establishments, residential areas, independent institutions, etc. If a decentralized treatment unit is installed in these areas, they may either generate excess reuse water, may need reuse water, or may consume the water that is already treated. Hence a complete assessment shall be made

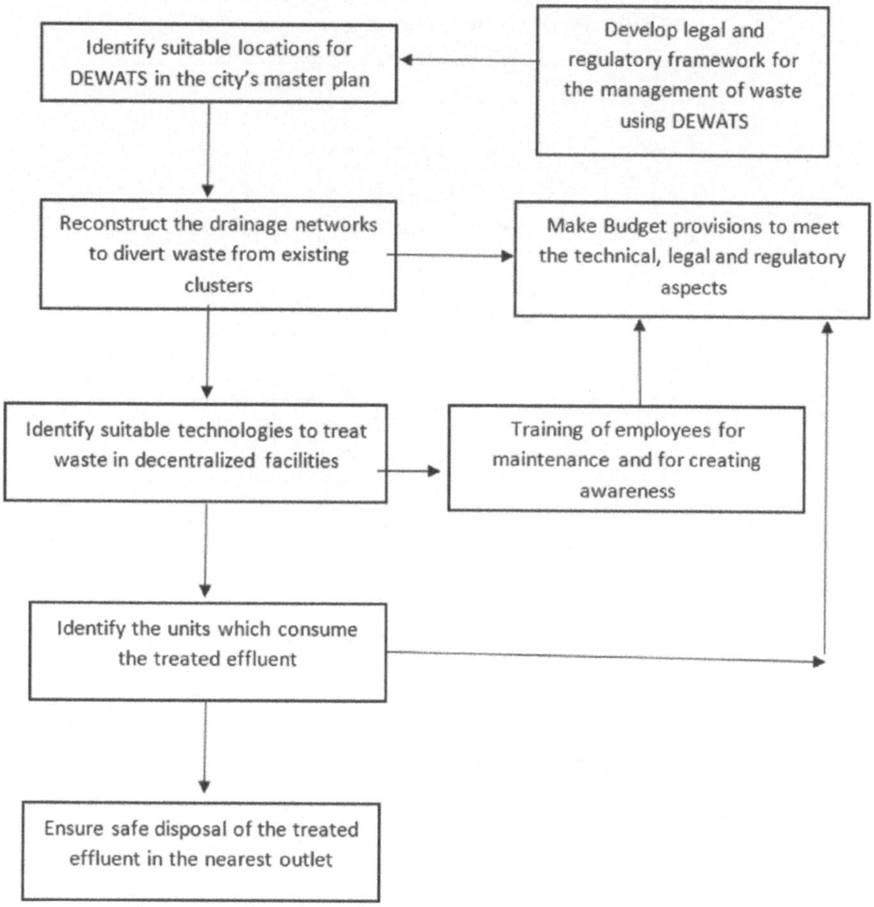

FIGURE 11.4 DEWATS system integration at the local level.

to identify the clusters that "balance" by reusing the effluent that is generated so that the overall efficiency of the system shall be improved.

5. The new areas that are going to be constructed should be connected to a decentralized system, and the cost of the construction of the decentralized system shall be uniformly distributed to the residents who are buying property in the new areas. The funds may be adjusted in the taxes they are going to pay to the municipal corporation/maintenance of the decentralized unit.
6. The separation of the urban runoff with the sewage generated shall be carried out in the initial stages to reduce the intake to the decentralized treatment systems.
7. In many instances, constructed wetlands are utilized to reduce the organic load at a place near to its generation. However, the systems are mostly damaged due to the sudden inflow of excess storm water damaging the system. Hence the decentralized treatment units shall be installed at a height above the maximum water level to reduce the impact due to urban flooding.
8. Indian railways have introduced bio-toilets in trains that can convert human waste into methane and carbon dioxide, and the liquid waste is disposed of after chlorination. If the same technique is replicated for handling waste in public toilets and other commercial areas with proper collection of methane, it will be possible to reduce the waste generation and help to generate methane as by-product.

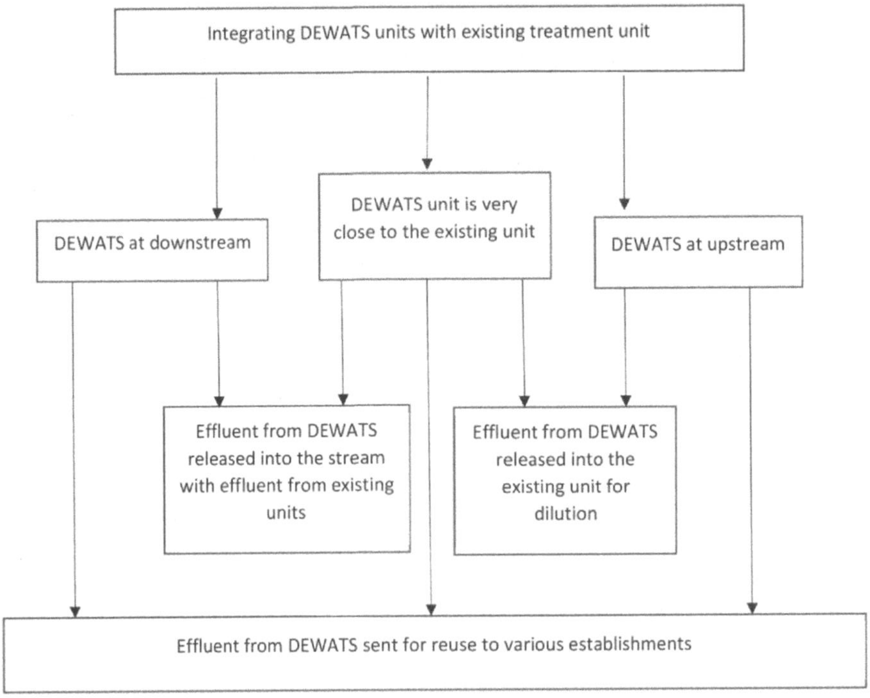

FIGURE 11.5 Integrated DEWATS systems with existing treatment units.

Decentralized wastewater treatments can be optimized in terms of effluent quality, treatment capacity, energy consumption, land consumption, land reuse, resource recovery, combination of treatment units and materials used, and cost of the project. Decentralized systems are recommended for reuse of wastewater at the point of generation itself. Treated water can be reused for sprinkling, flushing, vehicle washing, floor cleanings, and safe discharge into water bodies. Optimization in treatment method enhances the quality of water, which can also be used for agriculture and drinking purposes. Considering the increasing population, decentralized systems can be optimized for high organic and hydraulic flow rates. Most of the decentralized systems consumes less energy compared to the other methods of treating wastewater. In order to promote and utilize green energy, solar panels are recommended for further analysis. Land is another important parameter to be considered for installation of decentralized water treatment. People may not be ready to share their quality land for a treatment facility. The design of decentralized systems should provide the land for reuse purposes. Underground structures or treatment units are suggested for public acceptance in terms of land use, aesthetic view, and odour. Resource recovery is an essential for circular economy and sustainable development. Proper sludge management and resource recovery will minimize the project cost. Selection of treatment units and materials used in treatment (recycled) can reduce the cost of the project and also improve the quality of effluent (College of Agriculture & Life Sciences Cooperative Extension 2010; CPHEEO 2022). The strategy for integrating the DEWATS (Decentralized Waste Water Treatment Systems) technology is shown in Figure 11.5.

11.6 CONCLUSIONS AND PERSPECTIVES

Since many civil engineering projects have not integrated water and wastewater technologies and energy in an integrated manner, the effort to integrate such large-scale existing projects is going to be a major challenge. If the user lacks understanding of the process, and the management in the local area does not follow proper guidelines for maintenance and operation of the decentralized systems, the communities may not accept these decentralized systems. Effective water management is essential to meet the needs of increasing water demand with increase in population and future development. Considering the water pollution, climate change, and water scarcity, efficient and economical water management is required for sustainable development. Decentralized water treatment methods are effective and economical for wastewater treatment. Decentralized facilities can be easily established in rural or urban areas with limited land consumption compared to centralized treatment processes. The wastewater can be treated at the point of generation, which allows reuse of water without any distribution networks. Decentralized treatment facilities can minimize the water pollution at local and global levels by discharging treated effluent into the water bodies. The effect of water pollution on human health and aquatic life is significant. Considering the increasing population and emerging pollutants in the day-to-day environment, it becomes difficult for the existing treatment units to take the increasing wastewater load and the increased load may lead to plant failure. It is difficult to frequently change the treatment technology in a centralized facility according to emerging pollutants. In order to address the same, decentralized treatment at the source of waste generation could be a better option, and it becomes easy to design appropriate treatment technology according to the waste generated. Implementing a decentralized treatment facility at the source of waste generation can control the illegal dumping of waste into the environment. Effective monitoring of waste generation and treatment can be achieved. Further, waterborne diseases due to water pollution are minimized. Public health and environment are protected for sustainable development. The challenges include effective monitoring and maintenance, quality control, technical expertise in communities for maintenance, and scale and capacity. Decentralized systems need to be designed with flexibility for scaling up in the future with respect to increasing population. Adopting advanced technologies in water treatment can lead to effective, efficient, and affordable decentralized treatment, for example, sensors and automation can improve monitoring and maintenance. Continuous developments in technology will be needed for cost-effective decentralized treatment. In order to promote such technologies, awareness among the people is important. Government support and monitoring is required for its expansion. Effective policy-making is required for its long-time function. Optimization in treatment combination, materials for construction, energy consumption, and resource recovery will promote this technology in an effective manner in public domain.

REFERENCES

AquaTechSystems 2021. https://communitysewer.com/2021/01/04/benefits-of-decentralized-treatment/. [Accessed 30.10.2022].

British Columbia Small Water Systems 2022. https://smallwatersystemsbc.ca/centralized-vs-decentralized-water-treatment-systems. [Accessed 30.10.2022].

Chirisa, Innocent, Elmond Bandauko, Abraham Matamanda, and Gladys Mandisvika. "Decentralized domestic wastewater systems in developing countries: The case study of Harare (Zimbabwe)." *Applied Water Science* 7, no. 3 (2017): 1069–1078.

College of Agriculture & Life Sciences Cooperative Extension, The University of Arizona 2010. https://wrrc.arizona.edu/awr/w10/decentralized.[Accessed 30.10.2022].

CPHEEO 2022. Guidelines for decentralized wastewater management, Guidelines for Decentralized Wastewater Management.pdf (cpheeo.gov.in). [Accessed 13.10.2022].

CSE INDIA 2022. Centre for Science and Environment, https://www.cseindia.org/mount/home. [Accessed 14.10.2022].

Starkl, Markus, Josephine Anthony, Enrique Aymerich, Norbert Brunner, Caroline Chubilleau, Sukanya Das, Makarand M Ghangrekar, Absar Ahmad Kazmi, Ligy Philip, and Anju Singh. "Interpreting best available technologies more flexibly: A policy perspective for municipal wastewater management in India and other developing countries." *Environmental Impact Assessment Review* 71 (2018): 132–141.

Tchobanoglous, G., L. Ruppe, H. Leverenz, and J. Darby. "Decentralized wastewater management: Challenges and opportunities for the twenty-first century." *Water Science and Technology: Water Supply* 4, no. 1 (2004): 95–102.

USEPA 2022. United States Environmental Protection Agency, https://www.epa.gov/septic/benefits-decentralized-wastewater-treatment-systems. [Accessed 30.10.2022].

Index

Note: Page numbers in bold denote tables, page numbers in italics denote figures.

S

T

U

V

W

Z